SPACE AND SPATIAL COGNITION

All living creatures inscribe their activity in space. Human beings acquire knowledge of this space by traversing it, listening to verbal descriptions, and looking at maps, atlases, and digital media. We memorize routes, compare distances mentally, and retrieve our starting place after a long journey. *Space and Spatial Cognition* provides an up-to-date introduction to the elements of human navigation and the mental representation of our environment.

This book explores the mental capacities which enable us to create shortcuts, imagine new pathways, and thus demonstrate our adaptation to the environment. Using a multidisciplinary approach which draws on psychology, neuroscience, geography, architecture, and the visual arts, the author presents answers to a number of questions. Which mental capacities do people mobilize when confronted with space? Which brain functions do they implement? How do digital technologies extend these capacities? By presenting space at the crossroads of a number of disciplines, this volume reveals how each of them enhances our understanding of human behavior in space.

Space and Spatial Cognition provides a unique insight into all facets of spatial cognition, including spatial behavior, language, and future technologies. It will be the ideal companion for all students and researchers in the field.

Michel Denis is emeritus senior research scientist at the French National Center for Scientific Research and a member of the Laboratory of Computer Science for Mechanics and Engineer Sciences (LIMSI-CNRS) in Orsay (France). As a researcher in cognitive psychology, he mainly studied the processes of mental representation involved in visual imagery and spatial cognition. He is the author of about 200 scientific publications in international journals and books.

Michel Denis takes us on a journey to the fascinating empire of space and spatial cognition. His book offers an insightful up-to-date account of topics that researchers and practitioners have been struggling with over centuries. Philosophy, geography, architecture, psychology, linguistics, and computer science have made considerable progress in the field over the past fifty years. A single outstanding mind describes the perspectives on space from all these disciplines. The author excellently masters the task of bringing them together in a joint and very well written story. The book is highly inspiring for experts and novices in the field alike.

Christian Freksa,
University of Bremen, Germany

Our lives are spent interacting in space, and spatial thinking pervades imagery in the mind. This engaging volume explains how that happens, how the neural processes underlying navigation in space allow marvelous feats of the imagination. The sources for these insights and their implications are as broad as they are deep: neuroscience, geography, human–computer interaction, architecture, and the arts. *Space and Spatial Cognition* is a must-read for anyone interested in the workings of the human mind.

Barbara Tversky,
Stanford University & Columbia Teachers College, USA

SPACE AND SPATIAL COGNITION

A Multidisciplinary Perspective

Michel Denis

Routledge
Taylor & Francis Group
LONDON AND NEW YORK

First published 2018
by Routledge
2 Park Square, Milton Park, Abingdon, Oxon OX14 4RN

and by Routledge
605 Third Avenue, New York, NY 10017

Routledge is an imprint of the Taylor & Francis Group, an informa business

British Library Cataloguing in Publication Data
A catalogue record for this book is available from the British Library

Library of Congress Cataloging in Publication Data
A catalog record for this book has been requested

ISBN: 978-1-138-09832-9 (hbk)
ISBN: 978-1-138-09833-6 (pbk)
ISBN: 978-1-315-10380-8 (ebk)

Typeset in Bembo
by Swales & Willis Ltd, Exeter, Devon, UK

CONTENTS

FIGURES

Foreword

Space and spatial knowledge

There is no living creature that eludes the issue of its inscription in space and its moving across it. For all animal species, moving is an imperative condition for survival, a fact that justifies the study of spatial behavior being at the heart of life sciences. Research in this domain first draws on observation and involves a variety of behavioral measures. It extends with the formulation of hypotheses about the internal representations that people build of space and the identification of their brain infrastructure. Joint understanding of behavior and of the cognitive and neurobiological mechanisms that subtend it is the guiding thread of most research programs dedicated to this field. In any case, it is mine.

The area

The title "Space and Spatial Knowledge" is intended to stress that beyond the analysis of the properties of *space* as an object of study in its own right – as promoted by mathematicians, philosophers, geographers, and others – a major challenge consists in understanding how living organisms acquire *spatial knowledge* and build representations that form the basis of their behavior. For a scientist, it is by their behavior that people ultimately express their knowledge of space. Establishing correspondences between behavioral markers and cognitive processes is the essence of the approach to be illustrated here.

There are multiple "sizes" of space, ranging from proximal space as it is captured by sight, like the page displayed on my computer screen, the surface of my desk, or even the room environment that surrounds me, to large-size environments that can only be known by moving around and experiencing a succession of perspective takings, as is the case with cities, regions, continents. This variety of "scales" raises the issue of establishing whether the representation of different spaces calls for distinct cognitive processes or is governed by the same single set of cognitive capacities.

A related issue is to establish whether or not spatial cognition constitutes a special domain of cognition. Some approaches may reflect the inclination for interpreting spatial cognition as a "module" with its specific structures and processes. Alternatively, rather than being an isolated entity within the cognitive architecture, the representation of space is seen as a function interfaced with the whole domain of human cognition, first of all perceptual and sensorimotor functions, extending to memory, learning, and reasoning.

Space, as a research area, places demands on disciplines that make use of diverse methodologies. In recent years, it has proved to be a highly multidisciplinary domain. The representation of space offers a privileged example of a field of research where a high number of disciplines converge, with a special contribution from those adopting the perspective of cognitive science.

Not only is space a topic actively investigated by a number of scientific communities, but within each discipline there is an expression of curiosity and interest about what the other disciplines tell about it. For this reason, even a book written with an initial disciplinary perspective – in the present case, cognitive psychology – cannot omit addressing issues that are relevant for other disciplines. This book has been planned with the ambition of gathering those pieces of knowledge that are relevant for life sciences, but also social sciences, language sciences, and computational sciences.

The book

The book adopts a point of view that recognizes primary, although non-exclusive, value to knowledge collected by behavioral sciences. This knowledge has been enriched and illuminated by data stemming out of many other disciplines, in particular neuroscience, linguistics, and computer science. The idea here was not to produce a formal textbook, but instead to provide a selection of solid findings resulting from a variety of empirical approaches. To serve this objective, the book has been organized around five thematic packages (each containing three chapters).

As a point of departure, I thought it relevant to consider *space as an object in itself* (Part I). The issue of space has been thoroughly investigated by generations of philosophers. It is also an essential component of geometry, a part of mathematics dedicated to the study of figures in space. Geography, on its side, is guided by the objective of accounting for how territories designed by nature and eventually reshaped by human history are distributed across space. In the final chapter of Part I, architecture and pictorial creativity are directly confronted with problems related to the expression of space in the creation of visual forms.

Space is also to be seen as a set of *environments where action takes place* (Part II). It offers a context in which a variety of behavioral instantiations unfold, such as reproducing travel, planning a trip, discovering a shortcut, and many others. Such behaviors are submitted to analyses that lean on the use of specific methodological tools. These analyses substantiate the theoretical accounts of the cognitive processes underlying spatial behavior. This is the subject of the second part of the book, which documents the behavioral counterparts of critical concepts, such as spatial reference frames and cognitive maps.

As a natural continuation of this approach, Part III reviews the brain mechanisms responsible for spatial cognition and behavior. The development of neuroimaging techniques has expanded our knowledge of the functions and dysfunctions of the *spatial brain*. The analysis extends to the diseases that affect spatial cognition and navigation when *sensorimotor systems* are damaged. Special consideration is given to the situation faced by people suffering visual impairments and to the efforts put in the development of navigational aid systems.

Part IV deals with *communication about space*. Beyond cartographic representations, much of spatial information can be expressed through verbal and digital forms of communication. While the previous two parts deal with knowledge and practice of space through the prism of action and behavior, in Part IV we tackle the issue of space as approached by linguistics and semantics. The question is to assess how spatial information is expressed not only in an analog format (where space is represented by space, subject to a change of scale), but also through natural or digital languages.

Lastly, Part V deals with *computational sciences* and the broad range of *technologies* currently developed with the aim of building devices and exploiting spatial databases, with special attention to those intended to serve human action, such as wayfinding in unfamiliar environments. As in many other domains of human–machine communication, engineers are confronted with the issue of optimizing the adjustment of their systems to the users' cognitive capacities. New spaces opened by virtual technologies, as well as challenges met by research in robotics, will also be discussed in this fifth and final part.

When completing this scientific journey at the crossroads of a range of disciplines, an epilogue will give me the opportunity to convince readers of the virtue of *spatial thinking*.

* * *

While writing this book, I could not avoid the feeling that the breadth of the subject and the variety of entailed questions might justify a more ambitious academic volume, a tentative *Treatise of Sciences of Space*. As a matter of fact, edited books have been published in recent years in the United States, but all remained confined to psychological and neurocognitive approaches (see Postma & van der Ham, 2017; Shah & Miyake, 2005; Waller & Nadel, 2013). However, a multidisciplinary perspective is essential in this domain and obliges us to adopt an approach that goes beyond solely brain and behavior sciences. Special consideration is due to the multidisciplinary approach illustrated in a book on geographic information systems and how their design must take data related to human behavior into account (see Richter & Winter, 2014).

The present volume reviews theories and data collected by scientists who belong to a range of disciplines, taking care that none of the major questions related to space are left aside. Each chapter can be seen as a key opening a door on a conception or a particular facet of spatial cognition. It is my wish that the amplitude of the topics discussed and the coverage of disciplinary approaches that are practically never illustrated in books written by psychologists open stimulating perspectives. These questions are treated in the form of compact, coordinated chapters in which more than half of the cited works have been published since the beginning of the twenty-first century, and which summarize the essentials of current knowledge on space and spatial cognition.

MD
Paris, Orsay, Dinard, La Laguna
31 March 2017

PART I

Space as object of knowledge and object of practice

Humans experienced space long before producing any discourse about it and, *a fortiori*, before building any scientific discourse. For most living creatures equipped with a nervous system, the environment is apprehended by perceptual experience and locomotion. It provides a frame in which actions are inscribed and motor programs set up, to progressively become refined. Since probably quite early in phylogeny, the environment has been internalized in the form of representations, which offer a substratum for implementation of action. At a more advanced stage in the evolution of the nervous system, the environment and the actions taking place within it became a subject for communication between individuals. For a given individual, the capacity to undertake procedures intended to transfer to a congener's mind his/her own internal representation of an environment – to serve a specific behavioral episode – undoubtedly marks a crucial stage in evolution. For the human species, it is correlated with the capacity to generate discourse about space, build a conceptual representation of it, apply measurement onto it, and draw laws and principles. Lastly, when it comes into play, scientific thinking brings with it a set of instruments – both intellectual and technical – that make it possible to grant space the status of an object of science.

To start with, how can we characterize this object? Space imposes itself on anyone as an intuitive reality which provides a context for a variety of sensory and locomotor experiences. It nevertheless remains an abstract concept. Speaking of physical entities endowed with spatial extension is not an insignificant task. It consists of not only describing the ingredients that inhabit spaces open to our experience, but also accounting for the principles that regulate a spatialized world. It is a fact that empirical sciences that contribute to substantiating the concept of "spatial cognition" do not deal so much with *space* properly speaking, but rather with people's *behavior* within that space and the *knowledge* they build of it. They primarily take space as a venue for action. This approach is biased towards behavior and representations, and is used by scientists as an angle of attack in uncovering the properties of the object "space." It will be amply illustrated in this book. But beyond its full validity, this empirical approach does not exempt scientists from an intellectual effort to figure out the concept itself, in particular through the philosophical reflection that space has elicited throughout centuries. We will also have to consider the contribution from disciplines whose proper object is space, such as geography and cartography, but also sciences of the Earth and of the Universe. Lastly, we will examine the disciplines that develop space-related practices such as architecture and the visual arts.

Only later will we dedicate ourselves, in Part II and afterwards, to what ordinary people (neither scientists nor practitioners) *know* about space and how they display their knowledge through navigational performance, spatial reasoning, and spatial discourse.

1

PHILOSOPHICAL APPROACHES TO SPACE

The concept of space attracts everyone's intuition. A review of the philosophical approaches to this notion reveals, however, a variety of points of view. Interestingly, it will appear that a pronounced tendency has developed, which consists of taking the *human* into account in delineating the concept of space. Also, we will see that Euclidean geometry is grounded in a geometric intuition shared by all, and even considered by some theoreticians as universal.

From absolute to relative space

We owe to Newton the formulation of an *absolute* conception of space. For the author of *Principia Mathematica*, space is a reality in itself. It exists as a permanent entity, independently of the matter that it contains, and even if it does not contain any matter. This radical conception is related to Newton's commitment to the idea of an immanent world. It was vigorously disputed by Leibniz, his contemporary. This philosophical dissent added to other controversies which brought the two philosophers into conflict (on the theory of gravitation or the paternity of the discovery of infinitesimal calculus). Contrary to Newton, Leibniz argued for space as being essentially *relative* (in fact, a property that space shares with time). For him, there is no space without matter, and space can only be conceived and seized by our mind because there are objects filling it. Populated with objects of the physical world, space is to be understood as the *collection of spatial relations* among these objects. These relations are expressed in terms of distances and directions. In sum, the author of the *New Essays on Human Understanding* subscribes to the notion of a discretization of space, at the expense of the intuitive postulate of its continuity. For him, space is an abstract reality, for which he proposes a vivid metaphor: space is to be conceived as the relationships between members of a family. All the concerned people are related to one another, but their relationships cannot be conceived independently of the existence of these people.

One century later, Kant's metaphysics would support the idea that although space and time can be apprehended as objective realities of the world, both are unanalyzable intuitions, inaccessible to rational reasoning, and providing to the mind the organizing frame of every human experience. But if the intuition of time has its origin within the individual, the intuition of space is engendered by the external world. It remains that Kant's space, contrary to the view promoted by Leibniz, is homogeneous and continuous.

At the beginning of the twentieth century, Bergson contributed more than anyone to the reflection on time, which durably prevailed in philosophical thinking, while pointing out also the cross-links between the concepts of time and space, in particular in the passage of his *Time and Free Will: Essay on the Immediate Data of Consciousness* (1910) where he analyzed the spatial counterparts (the course of the hands of a clock) of temporal events (the time elapsed during that course). It is in this context that Bergson introduced the famous formula by which he characterizes time as "the ghost of space" (p. 99).

The notion of the continuity of space was still present in the views promoted by Franz Brentano while he was drawing the first outlines of his phenomenological approach.[1] Brentano did not postulate any radical opposition between the outer world and the psychological world. Both are associated by a relation of "intentionality," which means that thought cannot be dissociated from the objects to which it applies, and that perception is always perception *of something*. Therefore, psychic phenomena are essentially characterized in terms of their relation to an object. This is the essence of every mental act to be *relational* in nature. Brentano is probably the first thinker to have noticed that the various philosophical conceptions of space were dependent on the meaning attached to the terms that refer to space. For instance, the ancient Greeks did not have at their disposal any term equivalent to what we call "space." The Greek word "topos" corresponds more precisely to our notion of "place," with the possibility for a specific object to be attached to this place. This terminological determination is probably related to Aristotle's conception of space as non-infinite, but instead bounded by the heavenly sphere. On his part, Brentano promoted the idea that space, like time, does not exist by itself. Both can only be conceived in terms of the intentional relations in which they are involved. In this sense, Brentano's approach is consistent with Leibniz's "relative" conception of space (see Kavanaugh, 2008).

Poincaré, space, and geometry

In the early years of the twentieth century, the most advanced approach to the concept of space was proposed by Henri Poincaré, who aptly articulated his philosophical views to a deep reflection on geometry. The most remarkable facet of Poincaré's thought was to take into account the individual's *body* to circumscribe the concept of space. In *Science and Hypothesis* (1905, first published in French in 1902), Poincaré questioned the notion that the space in which our images of external objects are displayed would be the same as the notion of space for geometers and would have the very same properties. He recalled the essential properties of "geometrical space": it is continuous; it is infinite; it has three dimensions; it is homogeneous; and it is isotropic.

Poincaré developed an analysis intended to demonstrate that this set of characteristics is not systematically found in what he calls "representative space." First, if the visual images formed on the retina can be considered as continuous, they in fact possess only two dimensions, which distinguishes "purely visual space" from geometrical space. Second, all the points on the retina do not play the same role, which casts doubt on the idea that visual space would be homogeneous. Third, if sight enables people to estimate distances and therefore to perceive the third dimension, this perception might well be restricted to the impressions associated with the effort of accommodation and ocular convergence. These muscular sensations are distinct from the visual sensations that convey to the individual the knowledge of the first two dimensions. Thus, the third dimension cannot be considered as playing the same role as the other two, so that visual space can hardly be viewed as isotropic.[2] Lastly, the tactile experience and the muscular sensations linked to movements also contribute to engender our notion of space, while remaining remote from the notion of geometrical space.

Thus, for Poincaré, "representative space in its triple form – visual, tactile, and motor – differs essentially from geometrical space. It is neither homogeneous nor isotropic; we cannot even say that it is of three dimensions" (p. 56). The consequence that Poincaré draws from this analysis is that "representative space is only an image of geometrical space, an image deformed by a kind of perspective, and we can only represent to ourselves objects by making them obey the laws of this perspective" (p. 57). The innovative view introduced by Poincaré is that when we localize an object in a given point of space, "it simply means that we represent to ourselves the movements that must take place to reach that object" (p. 57). To represent to ourselves these movements means to represent to ourselves the muscular sensations that accompany them, but these sensations have obviously no geometrical character and do not imply the pre-existence of the notion of space.

The text on the relativity of space, first published in 1906 in *L'Année Psychologique*, then in *Science and Method* (1914), opens with a strong statement: "It is impossible to picture empty space" (p. 93). Poincaré rejects the concept of absolute space and maintains the principle of relativity of space, following on Leibniz's theory. He casts doubt on the idea that one would claim to "know" the distance between two points. According to him, any reference to "absolute space" is meaningless. There can be no question of absolute magnitude, but of the ratio of that magnitude to the instrument used to measure it (a meter, the path traversed by light, etc.). The notion of a "direct intuition of space" (intuition of distance, of direction, of a straight line) is illusory. If people ultimately succeed in "building up" a sense of space, it is thanks to an instrument that they use naturally, namely, their own body. "It is in reference to our own body that we locate exterior objects [. . .]. It is our body that serves us, so to speak, as a system of axes of co-ordinates" (p. 100). This approach applies to "restricted space" (nowadays, one would refer to "pericorporeal space") as well as "extended space" (the wider space accessed when one moves). In this space, points are defined by the succession of movements to be made in order to reach them, starting from the initial position of our body. Poincaré acknowledges that it is possible to translate our physics into the language of a geometry of more or less three dimensions, but points that "the language of three dimensions seems the best suited to the description of our world" (p. 113). If the contrast between the roughness of a "primitive geometry" and "the infinite precision of the geometry of geometricians" is well recognized, it remains true that "the latter is the child of the former" (pp. 114–115). But, of course, "it required to be fertilised by the faculty we have of constructing mathematical concepts, such, for instance, as that of the group" (p. 115). Therefore, geometry, which is not an experimental science, is nevertheless connected to experience. If we have created the space that geometry studies, it is by adapting it to the world in which we live.

In the historical context where Euclidean geometry was confronted with the "other" geometries (hyperbolic, elliptic, etc.), the question of the nature of real space was hotly debated. Poincaré's position consisted of proposing that the choice of a geometry to describe space is finally a matter of convention. Euclidean geometry is certainly "simpler" than a non-Euclidean one, which explains that it is preferred when we have to account for the geometry of our surrounding world. Simplicity and efficiency appear to be plausible criteria for the selection of a geometrical system at the expense of another. Here we realize the break endorsed by the author of *Science and Method*: intuition is recognized as contributing to the foundation of mathematics, including geometry. The human mind easily accesses the notion of continuity, in particular when it applies to the domain of space. Therefore, there is an intuitive grounding of geometry, and it is on this intuition that the mathematical construction of the discipline relies.

This idea is therefore the opposite to that of a geometrical space governed by universal logic rules, independent of humans. For instance, for Frege, these axioms can only be expressed in a highly abstract language and their objectivity is the very condition of the foundations of mathematics. But this position simply neglects that it is the human, as a living being in the world, who constructs mathematical concepts, and not any other thinking and acting entity. Geometry, initially the science of figures, then becoming the science of space, tended to become, thanks to Poincaré, the science of *movement* in space. This approach prefigured a broader, more recent view of science which takes into account the relation of the human to the world and the knowledge that the human constructs of it. For instance, as regards geometry, the exchanges with the experimental disciplines dedicated to the study of movement (gesture, locomotion) and spatial knowledge establish new connections between "mathematical geometry" and the "geometry of the sensible world" (see Longo, 2003). In these conditions,

> our practice of the invariance of objects of the world [. . .] allows us to build or conceive in retrospect the "invariance" or stability of our conscious representations, those of language and space, for example, and those most stable and invariant conceptual constructions, those of mathematics.
>
> *(p. 4)*

By promoting a "modern philosophy of nature," today's science entertains the idea that "the construction of concepts is rooted upon praxes originating with our humanity" (Bailly & Longo, 2011, p. 2). In the domain discussed here, the understanding of space relies on the functioning of neuronal structures which make it possible for the brain to give spatial and temporal coherence to the information coming from various sensory modalities (vision, audition, proprioception) (see Berthoz, 2000).

Poincaré's thinking appears to have influenced Albert Einstein at the time he was elaborating his theory of restricted relativity. It is documented that Einstein read *Science and Hypothesis*. In 1905, his article in *Annalen der Physik* on electrodynamics of moving bodies directly questioned the physics of Newton, which was still influential in the community of physicists. Einstein illustrated a radical rupture with the Newtonian postulate of absolute time and space, to the advantage of a relative conception thereof. His approach would be extended in 1916 by the formulation of the theory of general relativity.

The idea highlighted by Einstein was that of a world made of a space–time four-dimensional continuum. If space is a three-dimensional continuum in which it is possible to define the position of a point with three coordinates (provided, however, that this point is considered as unmoving), the events of the spatio-temporal world are seen as including an additional coordinate, the temporal one. This conception breaks with Newtonian mechanics, which considers time as a continuum independent of the spatial continuum. For Einstein, "time is robbed of its independence." The theory of relativity tightly couples it to space and establishes their mutual dependence (see Einstein, 1920).

Merleau-Ponty and phenomenology of space

The issue of space or, better, "spaces" (plural) is an important focus of the writings of Maurice Merleau-Ponty. In the period which immediately followed World War II, spatiality again took a prominent place in philosophical thought, after years where philosophy had given priority to the concept of time. Merleau-Ponty contributed to this perspective, as he was about to teach "pedagogical psychology" at the Sorbonne and as Piaget was launching his investigation of the child's representation of space.

First published in 1945, *Phenomenology of Perception* was translated into English in 1962. Merleau-Ponty contributed to the renewal of philosophical concern about space, while displaying an eminently personal point of view on this subject. He expressed much reservation with respect to Euclidean space, which is supposed to correspond to an "objective" form of space, but is for him nothing but "a space without transcendence," nothing more than "a network of straight lines." On the opposite, he claimed the primacy of subjective space and attempted to identify the features of what he called "original space," a concept connected to the idea of "primordial spatiality." He gave much consideration to the spatiality of an individual's body and to the "body scheme," whose spatiality cannot be equated to that of external objects. "If my arm is resting on the table I should never think of saying that it is *beside* the ash-tray in the way in which the ash-tray is beside the telephone. The outline of my body is a frontier which ordinary spatial relations do not cross" (p. 98).

Still more significantly, Merleau-Ponty rejected the notion of a universal space and highlighted the idea of *plurality* of spaces, suggesting that each has its particular philosophical status. For this reason, he considered space experienced by specific individuals and in specific contexts, such as space traversed by the child, dreamer's space, space of the schizophrenic, space of "primitive" people. There is no homogeneous and isotropic spatiality for each of these experiences, and the relations inscribed in space can only be related to the person who holds and describes them. Merleau-Ponty analyzed other forms of experience, such as those of painting, music, dance, and, in a particularly inspired passage, night. "Night has no outlines; [. . .] it is pure depth without foreground or background, without surfaces and without any distance separating it from me" (p. 283). Merleau-Ponty reviewed a variety of "qualitative spaces," positing that the condition for producing a philosophical discourse about spatiality requires to take into account "the establishment of the subject in a setting, and finally his inherence in a world" (p. 250).

Phenomenology thus assigned to itself the objective of apprehending the individual's actual experience. The view that this individual has on an "experienced space" has nothing to do with the geographer's look at a map of the world. The originary experience of space is a subjective one, directly involving the person's body. This is not a body among a set of objects distributed across space, separated from each other by various distances, and viewed from a perspective. The individual's body, as the basis of subjectivity, is recognized as the very principle of any spatial knowledge (see Vetö, 2008).

Phenomenology substitutes to "geometric distance" the concept of "experienced distance," with the consequence that for an object that might be located at different distances from a person, "the object a hundred yards away is not real and present in the sense in which it is at ten yards" (p. 302). They simply do not appear to be the same object. The notion of "size," then, is to be preferred to that of "distance" to account for this phenomenon.

> It is, therefore, quite true that any perception of a thing, a shape or a size as real, any perceptual constancy refers back to the positing of a world and of a system of experience in which my body is inescapably linked with phenomena. But the system of experience is not arrayed before me as if I were God, it is lived by me from a certain point of view; I am not the spectator, I am involved, and it is my involvement in a point of view which makes possible both the finiteness of my perception and its opening out upon the complete world as a horizon of every perception.
>
> *(pp. 303–304)*

The stress on the concept of "point of view" and the claim of the reductive character of "survey view" forecast the cognitive approaches which, fifty years later, would lead to the distinction between egocentric and allocentric perspectives in the building of spatial knowledge. Similarly, the criticism of the geometric view of the three dimensions (height, width, depth) as "substitutable" to each other was echoed by the empirical works of Franklin and Tversky (1990), who empirically demonstrated their unequal cognitive accessibility. For Merleau-Ponty, if all dimensions are "existential," depth is still more so than the other two and has a predominant role in the building of spatial knowledge.

Interestingly, Merleau-Ponty's philosophical approach borrowed concepts from topology, for instance through repeated developments about the notion of "impingement." The author even suggested that philosophical thought is itself the expression of some spatiality. Impingement, intertwining of concepts, transitions between domains are spatial metaphorical features of the philosophical act itself. Philosophy might be inherently "spatialized," due to the body/world relations that it aims at accounting for (see Saint Aubert, 2004).

Genetic epistemology of spatial representation

The common postulate of modern theories of space is that one cannot speak about space without referring to the individual and the individual's actions within it. However, reference is typically made to a "theoretical" individual, with only modest, if not absent, consideration of the changes that may affect the individual's capacities along his/her development. This aspect was in fact at the core of Jean Piaget's "psychogenetic" theory, the opposite extreme of a nativist conception of knowledge. Piaget focused on the individuals and their activities. The objects in space do have significant properties, but still more important are the *actions* that people carry out on these objects. The objective is then to answer a set of specific research questions: Through which stages will children build their knowledge of space? How do children acquire the capacity of measuring lengths? How do these early acquisitions result in later representations characterized by quantification and logic? These questions motivated the works reported in the two major books originally published in 1948 (Piaget & Inhelder, 1967; Piaget, Inhelder, & Szeminska, 1960). The observations collected from children aged 2–12 conferred empirical validity to a research program that was anchored in the philosophical reflections of the early twentieth century. This approach provided an ambitious dimension to the Piagetian project of establishing genetic epistemology, a general explanatory model of knowledge acquisition.

At the beginning is sensorimotor space, which is directly linked to the actions and perceptions of the child. It is constructed from manipulations and moves of the individual. This space is in fact incomplete because it remains connected to the child's close environment under its various aspects, without any possibility for such fragments to be unified into a single representation. The availability of sensorimotor space will progressively open such a possibility, although an integrated representation of space as a whole will still be lacking. This "action space" is defined by the four properties of the practical group of displacements: identity, composition, reversibility, and associativity (the latter one being involved in the generation of detour behavior).

The transition to more advanced forms of spatial knowledge depends on the creation of a "representational space," first characterized by static imaginal representations and the difficulty for the child to represent transformations. It is only at the preoperational stage that objects will be identified in terms of their *topological* identity, with a spatial intuition governed by relations of neighborhood, separation, order, covering, and continuity. These relations remain internal to the figures, and are not used to establish relations between figures. They only allow the child to refer to his/her own point of view, which means that topological space has not yet the properties of a "total space" that would encompass all figures.

The situation changes radically when the child accesses *projective space*, when figures are no longer considered only for themselves, but can be envisaged by reference to a "point of view." The child now envisions that there may be other points of view than his/her own on an environment. When placed in front of a scene (as the famous scenery of the "three mountains"), the child is now capable of imagining the representation that another observer may construct from the opposite point of view. The capacity of putting oneself in another person's place and the decentration rendered possible when one gets free from the egocentric vision of an environment are decisive steps toward the construction of total space. These steps can only be taken when transformational operations (such as rotation and inversion), which are characteristic of the operational stage, are effectively mastered. Only then does the child reach the notion of a "complete relativity of perspectives."

Whereas projective space rests upon the coordination of points of view and figures seen from these points, access to *Euclidean space* makes it possible to posit objects (including oneself, as an object) in a coordinate system. Euclidean space not only coordinates the positions of objects, but also their metrical relations. The whole space becomes the single frame of all the objects and all their (real or potential) observers. When reaching the formal stage, the adolescent becomes able to create representations of that space, whose metric properties can be measured in an objective way. The person then deals with a space that sustains mathematical quantification.

Universality of geometric intuition?

Piaget's work provides the most solid argumentation in support of the psychological foundations of geometry. It is in continuity with Poincaré's postulate that the mathematical construction of geometry relies on contents provided by human intuition. This view has continued to spread in recent years, suggesting that the system on the basis of which geometry developed is the cognitive system of spatial representation that is available to human beings. A further notion is that of the universality of the cognitive "modules" that have subtended the foundation of Euclidean geometry.

Arguments in favor of this postulate are provided by the works which attest for the sensitivity of people without any familiarity to geometry to some properties of geometric objects. The most remarkable example is that of the Mundurukú, an isolated Amazonian population without any contact with geometric concepts and whose language does not include any terms to designate classic geometric concepts (such as parallelism or right angle). The princeps experiment involved children and adults from this population. Sets of six figures each were presented to them, each set illustrating a geometric property (for instance, the parallelism of two lines) in five of these figures, while one figure did not display this property (Figure 1.1). The participants were asked to designate which of the six figures was "very different" or "weird" in comparison to the others. The results showed that the Mundurukú identified quite easily the figure that lacked the property common to the other five, with a performance level well above chance. Furthermore, their performance was found to be not different from North American participants, who had received formal geometric education. Lastly, out of the forty tasks to which the participants were submitted, and which were hierarchized in terms of difficulty, the tasks ranked in the same order for the Mundurukú and the North Americans. The parallel ranking of task difficulty argues for similar geometric intuition in both cultures, independent of knowledge acquired via teaching of geometry (see Dehaene, Izard, Pica, & Spelke, 2006).

Clearly converging are the data that show equal sensitivity of the two cultures in detecting differences in lengths or angles, two essential ingredients of geometric figures. These two dimensions obviously have a privileged status in the perception of shapes, contrary to figure orientation, which is a non-constraining feature in the context of transformational geometry. It turns out that the observers' sensitivity to lengths and angles is confirmed, while it is not the case for figure orientation, and this is true for both populations. In other words, a kind of universal geometric intuition would make it possible for people to capture relevant invariants, namely, those which ensure the identification and classification of spatial objects (see Izard, Pica, Dehaene, Hinchey, & Spelke, 2011; Spelke, Lee, & Izard, 2010).

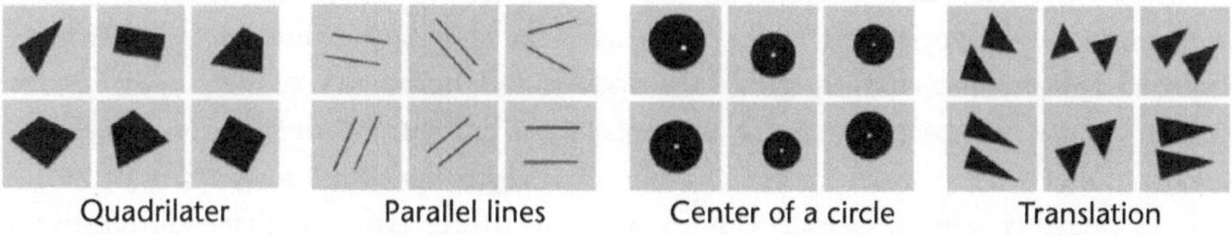

FIGURE 1.1 Four series of six geometric figures. Adapted from the chapter by V. Izard, P. Pica, S. Dehaene, D. Hinchey, and E. Spelke in *Space, Time and Number in the Brain: Searching for the Foundations of Mathematical Thought* (2011), by permission of V. Izard, Elsevier, and the American Association for the Advancement of Science

Comparative approaches in the domain of space and geometry thus attest that the same cognitive competences are shared by people from different cultures regardless of their different experiences in their respective environments. Furthermore, they inform us on *geometry* itself, by attesting that Euclidean geometry is not a pure mental construction proposed by specialists of the domain, which could only benefit people who have received special training. Strong arguments, on the contrary, emerge in favor of the universality of the ingredients of Euclidean geometry within the cognitive systems that allow people to perceive and represent space. There is no doubt that geometric knowledge, as demonstrated by people who remained well away from any formal education on this matter, can be enriched and refined by culture (through the acquisition of specialized competences pertaining to the use of mathematical or cartographic tools). Research initially anchored in philosophical thinking, then supported by cognitive sciences, provides light on genuine geometric knowledge inherent in the human mind. This connects back with Kant's metaphysics and the notion of natural, immediately available knowledge of space, notably geometric space, for the human mind.

Numbers and space

While geometry is explicitly dedicated to space, other areas of mathematics are concerned as well. This is the case of numbers. Numbers express quantities which have no intrinsic spatial extent in principle. Nevertheless, a broad set of data converge to suggest that they are systematically associated with space.

Pythagoras, of course, is the first reference on this matter. By postulating the existence of a close interface between arithmetic and geometry, the Pythagorean doctrine promoted the value of developing a geometric representation of numbers. This was illustrated, in particular, by the design of geometric objects, such as the "tetractys," an equilateral triangle whose unique feature is to contain the spatialized expression of the first four natural numbers (Figure 1.2). On such grounds, geometric figures are expected to serve as media to support arithmetic demonstration.

In the very early days of modern psychology, the possibility for numerical information to be mapped into representational space was first illustrated by Francis Galton (1880), who documented "the innate and hereditary tendency of certain persons to see numbers in definite and constant arrangements or schemes" (p. 253). Galton collected reports from people who claimed to spontaneously think of numbers as occupying precise locations in space, forming two- or three-dimensional objects. When exposed to auditory presentation of numbers, such people automatically "see" them in a structured mental space. Some of them create visual patterns in which numerical information is displayed in particular complex configurations. The occurrence of such mental representations of numbers that display vivid visuo-spatial properties was confirmed by a number of further observations (see Seron, Pesenti, Noël, Deloche, & Cornet, 1992).

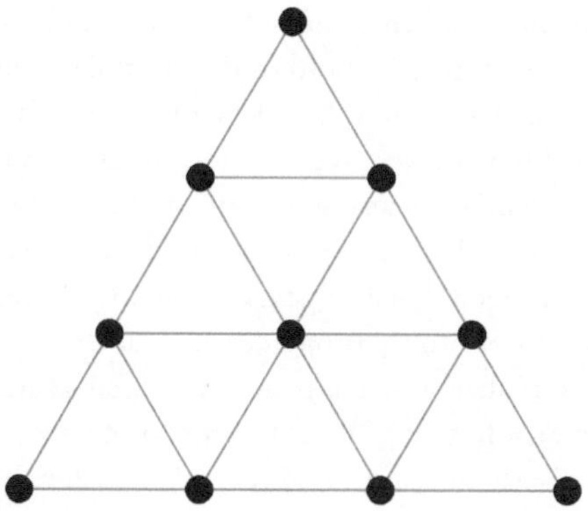

FIGURE 1.2 Tetractys

Most of the reported cases in this area are those of somewhat "exceptional" people, but the connection between numbers and space is attested more broadly. Indeed, mathematical thinking is open to the notion that the mental representation of numbers depends on spatial representations, and that numerical magnitudes are represented internally by a spatial extension. In particular, Dehaene (1997) promoted the view that the representation of number meaning is organized spatially, mapped onto an analog "mental number line." Quite compelling are the behavioral data attesting that the presentation of digits automatically draws people's attention to specific parts of their visual field and, as a result, enhances motor responses associated with corresponding portions of space. In sum, the cognitive system is thought to map the numerical continuum onto an internal representation of space, which in turn measurably affects motor performance.

A demonstration of this phenomenon was obtained by Dehaene and his colleagues in a classification task where people were presented with Arabic digits in the range 0–9 and asked to decide, by pressing one of two response keys (with their left or right hand), whether the target was odd or even. The results showed that people responded faster to relatively small numbers with the left hand than with the right hand, and faster to relatively large numbers with the right hand. The spatial congruence between the response side and the relative position of the represented numerical quantity on a mental number line strongly suggests that the numerical representation accessed in the task contains genuine spatial information. This phenomenon is known as the spatial-numerical association of response codes (SNARC) effect (see Dehaene, Bossini, & Giraux, 1993). The evidence is indicative of the fact that numerical magnitude is processed automatically, and that the effect takes place without any conscious experience on the part of the people who show it (see Wood, Willmes, Nuerk, & Fischer, 2008).

The association of numbers with visuo-spatial representations is well documented in the general population, and more particularly in those people who claim to possess rich visuo-spatial representations of numbers and rely on them to maintain numerical information in an active state. This is true in particular of calculating prodigies, who typically hold and manipulate numbers onto a visual short-term representational medium (see Pesenti et al., 2001). Noticeably, the neural circuitry responsible for the representation of quantity is located in the parietal lobe, a region that overlaps with the neural network involved in the representation of space (see Hubbard, Piazza, Pinel, & Dehaene, 2005).

The SNARC effect reflects what appears to be an inherent spatial property of the internal representation of number magnitude. But the spontaneous mapping of information into representational space is not restricted to numerical information. Suggestive hints are available to the fact that non-numerical magnitude may also be spatially coded. This is the case of letters of the alphabet or months of the year, which, like numbers, are structured along ordered sequences. For instance, when people are asked to judge the position of months as coming before or after July, months from the beginning of the year are responded to faster with the left hand than with the right hand, whereas the reverse pattern is obtained for months towards the end of the year (see Gevers, Reynvoet, & Fias, 2003).

Suggestions have been made that space and quantity might be part of a generalized magnitude system. Space, quantity, and time are linked by a common metric, and operate on similar and partly shared principles. As regards the brain networks governing these principles, Walsh (2003) has suggested the inferior parietal cortex to be the most likely locus of a common magnitude system. The SNARC effect might well prove to be a particular case of a broader spatial-quantity association of response codes (SQUARC) effect.

In line with the findings documenting the SNARC effect, variations on the original paradigm provided further support to the fact that the perception of numbers automatically activates internal spatial representations. One of them consists of presenting people with a horizontal line flanked by numbers (for instance, "3" is printed at one end of the line and "9" at the opposite end). People are then asked to perform a very simple bisection task, that is, to mark with a pen the center of the line. The results show that bisection is modulated by the numerical information, independent of any explicit processing of numbers (which, by the way, are irrelevant to the task). Nevertheless, people systematically make a bisection error towards the larger magnitude number. When the larger digit is the left-hand sided flanker, a leftward bias emerges, whereas a rightward bias is evidenced when the larger digit is the right-hand sided flanker. This provides further support to the notion of automatic association of number magnitude with spatial representations and response codes (see de Hevia, Girelli, & Vallar, 2006; Fischer, 2001).

There are good reasons to expect that the orientation of the numerical continuum is shaped by cultural constraints. In people from cultures who read and write from right to left in their native language, the SNARC effect is reversed. Presumably, the spatial properties attached to numbers are acquired by children during reading and writing acquisition. However, by using experimental procedures adapted to young children, de Hevia and Spelke (2009) showed that the mental number line found in adults emerges much earlier in development than expected. They defended the view that space and number are intrinsically related by a common mental metric, and that this association is not the plain result of a cultural construction. According to this view, numerical and spatial representations are assumed to be linked prior to the onset of formal instruction, in a manner that suggests a fundamental link between the domains of number and space.

Besides numerical magnitude, another continuum that entertains potential relations with spatial representations is pitch height. In many languages, the classification of pitch relies on spatial terms, such as "high" or "low," according to associations which in fact do not appear to be arbitrary. In an experiment by Elena Rusconi and her colleagues, people were invited to compare the frequency of two pure tones. After listening to a reference tone, a second tone was presented, and people had simply to decide whether the probe tone was higher or lower than the reference one. To this effect, they had to press response keys which were located in the upper and the lower part of a keyboard. It turned out that a clear advantage appeared for "up" responses to high-frequency and "down" responses to low-frequency pitches (compatible mapping), as compared to "down" responses to high-frequency and "up" responses to low-frequency pitches (incompatible mapping). Performance was faster and more accurate when participants provided responses that were vertically aligned to pitch height. This effect was still observed when instructions were free of any spatial words and did not lead people to focus on the concept of height (in a task where people had to simply identify the timbres of various instruments) (see Rusconi, Kwan, Giordano, Umiltà, & Butterworth, 2006).

The authors proposed referring to this phenomenon as a spatial-musical association of response codes (SMARC) effect. It was confirmed with trained musicians, who also exhibited a *horizontal* SMARC effect. Musicians (not only pianists) are familiar with keyboards on which low-frequency pitches are generated by pressing leftward keys and high-frequency pitches are played by rightward keys. Interestingly, musical training seems to elicit a specific vertical-to-horizontal relationship, with preferential mapping of low pitches on left-side responses and high pitches on right-side responses. Beyond the attested analogy between music perception and mathematical cognition, it is remarkable that both domains entertain special relationships with the mental representation of space.[3]

Notes

1 The texts most directly related to the issue of space, as dictated by Brentano between 1913 and 1917, were collected and edited by S. Körner and R. M. Chisholm in 1976 under the title *Philosophische Untersuchungen zu Raum, Zeit und Kontinuum*. The volume was then translated into English by B. Smith (see Brentano, 1988).

2 In Chapter IX of Benjamin Bourdon's volume, *La Perception Visuelle de l'Espace* (*Visual Perception of Space*), published the same year as the first edition of Poincaré's book (1902), one finds the same reference to the fact that the sensations that accompany changes of convergence would be responsible for the perception of depth.

3 The SMARC effect is only one of the many aspects by which music brings to light its connections with space. The graphic systems used for music notation reflect a variety of solutions to encode sound frequencies and durations spatially in musical scores. The left/right distribution of sounds across keyboard instruments probably conforms to non-arbitrary regularities. Lastly, symphonic orchestras display groups of instruments on the stage according to traditional arrangements, which nevertheless remain open to innovative revisions when a composer or a director wants to renew the spatial distribution of musical sounds (see Xenakis, 1971).

2

GEOGRAPHIC SPACE

Geography is often seen as the science of space *par excellence* (see Thrift, 2003). It deals with territories distributed over the surface of the earth, from a variety of perspectives, such as the nature of soil, its morphology, the way in which populations occupy space, the modes of living and political regimes. Whether the focus is put on the characteristics of territories or their inhabitants, geography remains concerned by objects that are distributed across a spatialized physical reality. Even if the spatial extension of this reality is not always at the core of geographers' discourse, their discipline etymologically remains "the science of earth."

Spatial environments, their contours, the shapes and figures carved by nature are the raw materials of which researchers develop a scientific analysis, which typically involves *measures*. The building of geographic knowledge relies on a methodological apparatus, which comprises instruments, computational methods, observation grids, etc., which allow geographers not only to collect, but also to interpret information, to extract meaning and build representations that can be shared with other scientists or users of that knowledge. For instance, if a natural form (such as the winding course of a river or the coast line of an island) must be expressed in a graphical form, the application of rational knowledge and the implementation of mathematical instrumentation will make it possible to put the data in order and confer greater legibility to a reality too complex to be kept in its original form in a representation. Most geographic objects are bound to be mathematically measured. Even though geography does not aim at quantifying every material it deals with, the understanding of increasingly complex processes, in which geophysical, biological, and social factors interact and call for interdisciplinary approach, it does call more and more upon mathematical modeling and computing resources.

Geographers' objects

Geographers are entitled to inquire into space as an abstract concept, in line with the philosophical questioning discussed in the previous chapter. They are also dedicated to the description of a range of spaces, as defined by a variety of features – geophysical, biological, climatic, human, political, and others. By doing so, they refer to "places," more or less extended, more or less precisely delimited, whose borders make it possible to distinguish between what is "within" that space and what is "outside." As scholars who have to cope with the concept of space, geographers will confirm the basic classification between the "in" and "out" concepts, even if demarcations sometimes remain fuzzy. However, when looking into spaces, geographers are led to take different levels of analysis into account, which in fact reflect the scientific programs of distinct communities within the discipline.

At the first level, space is considered in terms of physically extended portions of a territory, amenable to quantification and measure. At a further level, the geographer no longer considers the territory as a physical surface analyzable on its own, but as an environment within which humans develop activities and social relations. At the third level, space is seen as an object of cultural representation, an object demanding social reflection on the world where humans' lives take place. This latter level has the value of revealing the variety of relations that human groups entertain with their surrounding space. This is the level considered by the French geographer Michel Lussault, who posits that "the social world is formed *via space* as a world of shared experiences for individuals and groups" (Lussault, 2007, p. 30).

Initially, the vocation of geographers was to build knowledge of territories and develop mapmaking. This used to be the quasi-exclusive role of geography from Antiquity to the Renaissance, with the creation of maps, atlases, portolan charts, which all consist in documenting remote spatial realities, but also assist navigation through these territories as well as maritime and coastal spaces. Later on, geography opened itself to other, wider and more ambitious questions, and started developing sophisticated cartographic projection systems, but its dedication to the description of territories never bypassed the *human* dimension attached to them. For instance, in the first century of the Christian era, while Strabo was devoting his *Geography* to the description of the parts of the world known by Greeks at that time, he promoted the idea that the character of people living in a particular region was determined by the properties of their spatial environment. The geographic position of a land in a continent (for instance, the fact that it has a maritime façade) is seen as a factor which contributes to opening up a population towards other territories, therefore to enhancing its fame and influence within and beyond this land. Strabo thus extended the notion of territory to that of "milieu" and introduced the idea that the features of a geographic environment may affect individuals' psychology and behavior. Centuries later, the "climate theory" (already found in Aristotle's works) was spelled out still more insistently by Montesquieu and Buffon, promoting the hypothesis of direct influence of environmental properties of a part of the world on the character of its inhabitants.

There are several ways for geographers to take the human dimension into account. Territories can be studied with the help of descriptors of their population (density, frequency of a given profession, of a given type of behavior, etc.). Here, population is treated as a parameter that characterizes the territory, of which it is just one of the attributes. But there is also a geographic approach directed to human beings as holders of values and representations, which aims at accounting for images that people build of their environment. Illustrations are available on the part of researchers working at the crossroads of social and environmental geography. In this spirit, there have been many attempts to collect graphic or verbal externalizations of "mental maps," this offering the possibility to compare the representations that various people construct of the same geographic space. Geographers have also provided evidence of the biases that affect spatial representations built by different groups of people living in the same environment. For instance, people living in a city of a multicultural country tend to underestimate kilometric distances that separate them from another city that belongs to their own community, whereas they overestimate distances to cities whose inhabitants mainly belong to another community, as was demonstrated in occupied territories of the Middle East where the relations between communities are conflictual (see Portugali, 1993).

Human geography has developed an increasing interest for "space as experience," for people's conceptions of their geographic and social environment, for the representations that they share as members of a community (see Frémont, 2005). The concept of space even comes to split up and the need appears to distinguish among the various places of living of an individual – residential spaces, working spaces, leisure spaces, etc. – each of which having its properties and specific operating modes. Lastly, geography is concerned with space and physical environments when people set about transforming them. The analysis of geographic spaces is thus guided by preoccupations linked to anthropization and its resulting changes (some transient, some long-lasting).

Because geography is interfaced with both social and behavioral sciences in its questioning about the cognitive relation of humans to space, it has come to entertain relationships with the broader area of cognitive sciences. To be clear, as is the case for other disciplines (such as linguistics or philosophy), only a limited sector of geography has developed such dialog with cognitive sciences. But, quite fruitfully, the connections of geography and cognitive psychology have resulted in benefits for the latter. An example is the use of mathematical tools which provide "visualizations" of mental representations of space, such as the topography of urban environments (see Cauvin & Schneider, 1989; Giraudo & Pailhous, 1994).

Spatial analysis and the concept of territory

Geography is committed to philosophical questioning about its object. An interesting contribution was provided by Immanuel Kant, who was also strongly committed to explicitation of geographic knowledge. His lectures at the University of Königsberg, first published in 1802, were intended to "systematically prepare youth to acquire knowledge about the world." The descriptive point of view was well illustrated. However, beyond the geographer's dedication to classification and inventory, Kant underscored the essential role of "reasoning" and knowledge of the "regulating principles" to be applied to the objects of geography (see Elden & Mendieta, 2011).

Today's geographers are focused on various forms of epistemological questioning. Is space viewed by a geographer the same as the space to which Kant, Poincaré and Merleau-Ponty dedicated themselves? It is sometimes noted that what is available to geographers and is subject to their analyses are just points, lines, and surfaces, all primitive elements whose combination creates delimitations within a physical space. Based on this observation, a conception of space has developed as an articulated set of places, distances, areas, likely to be represented graphically, and also to undergo mathematical formalization (see Brunet, 1987). At the same time, there is a growing expression of interest for the concept of "occupied space" (occupied by geophysical realities, by human groups, by human-made constructions, by circulation networks). This concern confers high relevance to the maxim frequently quoted in the geographic community: "Space only exists by what fills it" (Moles & Rohmer, 1998, p. 38).

In the last few decades, the concept of "spatial analysis" became quite popular in the context of a broad ambition to account for any sort of spatially inscribed phenomenon. The analysis of phenomena in terms of their spatial organization is put at the service of the comprehension of their properties and the search for explanatory models. The foremost example in this domain is the cartography of cholera cases during the epidemic that affected the population of London in 1854. By examining the spatial distribution of identified cases, John Snow, a physician in charge of the situation, was able to localize the source of the outbreak, which was spread through the water distribution network (Figure 2.1). According to this perspective, geographic space is no longer interpreted as a plain surface or support, but the revealing factor of structures within which the covariation of quantified properties provides unique information to the researcher. This approach is relevant for many disciplines, among which economics is at the forefront, as illustrated by the development of "spatial econometrics." A trend has also appeared to take into account spatial structures in a purely abstract perspective, without any reference to the processes that generate these structures (see Getis & Paelinck, 2004).

The development of geographic information systems (GISs) offers abundant packages of data to researchers dedicated to spatial analysis. The concepts of distance, connectivity, and directional relations between places have all become concepts of major significance. This approach also profits from the increasing capacities of visualization offered by these systems. Visualization thus must be "guided" to carry full information, especially for users who are not geographers, in such a way that a clear-cut distinction is recognized between "data," which result from the capture and quantification of observed phenomena, and "information," which results from reasoning and interpretation. This distinction is all the more important given that databases regarding land use are widely taken into account for decision-making in territory development and planning.

The notion of territory has gained importance in today's geography. The ethological meaning of the term is well attested when it refers to the habitat area where an animal population lives and reproduces. For human and political geography, space becomes territory as soon as common practices, identity sharing, and control

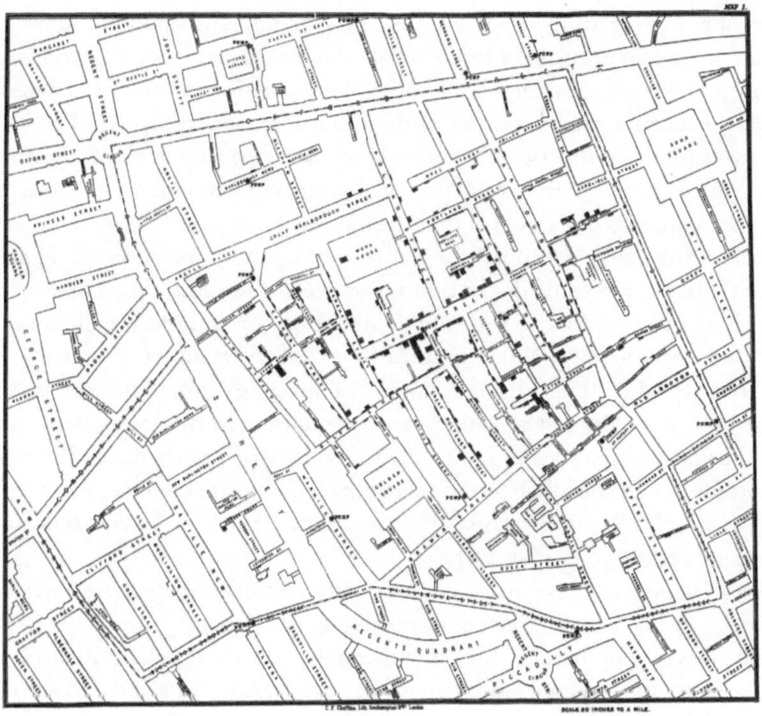

FIGURE 2.1 Cartography of cholera cases in the London epidemic of 1854

strategies regulate the actors' activities. A territory is a portion of space marked by specific features (natural, administrative, political, cultural, linguistic, etc.), limited by borders, but its meaning obviously stands well beyond a purely geographic frame. The concept includes a dimension of social identity, in such a way that the spatial reality circumscribed by the territory is supplemented by the symbolic value attached to the human group that lives there and the actions that contributed to forging its historic identity. For Claude Raffestin (1980), promoter of the concept of territoriality, a territory is more than space. It is a space enriched with meaning, with notions of power and control attached to it. "A territory is to be considered as space informed by the semiosphere" (Raffestin, 1986, p. 177).

The issue of territoriality has inspired a number of works, of which only a few can be mentioned here. A number of them are aimed at bringing out the properties of entities identified as territories. A territory refers to the materiality of a geographic space, but also bears a symbolic sense which allows people to experience themselves as *belonging* to this territory. Between two territories or two portions of a more extended space, there is discontinuity, but also the notion of contact between them. Discontinuity may be assessed by quantitative or qualitative differences. It can be linear, when the two portions of space are neatly differentiated (in the form of a border line), but it can also be thicker, when discontinuity is characterized by some gradient (in the case of a no man's land). But beyond the notion of border or limit, as physical reality signing the discontinuity between two territories, the two portions of space can also be seen as achieving an *interface*, due to the contact between two neighbor entities. The analysis in terms of interface makes it easier to account for the notion of flows between the territories. The interface plays two important functions in geographic space, the functions of exchange and regulation, while entertaining the potential perspective of interaction between the territories, and, in some cases, of their fusion (see Chapelon & Emsellem, 2008).

Other works have extended the concept of territory by introducing the possibility for territories to be discontinuous or even "multilocated." Denise Pumain (2016) has envisaged "many forms of delimitation (or boundary-making) ranging from inclusion within national boundaries [. . .] to fragmented units [. . .] and networks" (p. 45). The classic example of splitting into multilocated territories is the case of diasporas. Other cases are illustrated by South African cities, where spatial practices inherited from the politics of apartheid imply traversing discontinuous and socially hermetic territories. These are complex forms of territoriality, which rely on network structures rather than continuous and bordered geographic areas (see Giraut, 2013).

Researchers at the crossroads of geography and demography are interested in the emergence of people's feeling of belonging to a territory, especially in today's context where populations are more mobile, nationally and internationally. Surveys have documented how people describe territories as being part of their identity heritage. They attest the prevalence of the local level as personal territorial reference (in France, the level of municipality rather than the level of department or region). Some regions, however, enjoy a high identity value, such as Brittany, Alsace, or Corsica. One notes also that for people who have histories of exile or migration and have experienced piecemeal identities, the attachment to a place is valued as the symbol of successful integration (see Guérin-Pace, 2006).

Maps

The Ancients' geography, which was primarily dedicated to the design of maps of the world, progressively evolved towards issues related to territories and the human activities that develop therein. Nevertheless, today's science of cartography faces serious challenges, in particular those related to the problems of representation inherent in the design of objects expected to configure geographic space. Cartographers are required to find appropriate means to express validly in two dimensions on a plane surface the convexity of geographic realities, especially when the representation encompasses large portions of the sphere of the earth.

The challenge was taken up in past centuries with a succession of inventive attempts. From the Mercator to the Winkel Tripel projections, cartography strived to find the least bad compromises between the distortions imposed on distances and those affecting the shape of continents. Today, GISs achieve representational feats through applications to an infinite variety of data (climatic, demographic, economic, sociological, etc.). They accommodate and make available to users information about places, buildings, services, and resources, while providing fine-grained localizations and advanced facilities for online processing. It is worth pointing out that all those sophisticated devices, which have become familiar to most users, are ultimately based on the processing of a limited number of geometric primitives: points, lines, and polygons.

In their daily, non-scholarly use, maps are thought to be the quintessence of analog representations, that is, *spatialized* representations of *spatial* realities. Quite simply, space is represented by space, with spatial relations being depicted homothetically in the form of spatial relations. Of course, spatial data can also be expressed under non-spatial forms. For instance, a chart displaying the two-by-two distances between the cities of a country contains information that is directly derived from measurements collected on the ground. Even so, it does not provide a representation from which human sight could capture the positional relations between cities. The recognized virtue of analog representations is to offer to an individual a substitute of real life to which the same perceptual and cognitive mechanisms would apply, the only difference being between the respective absolute sizes of the representation and of the represented object. From a biological perspective, it is advantageous for an organism to reuse mechanisms that are available and whose functional value has already been experienced.

Maps entertain with represented territories a relation of structural similarity, which makes it possible to plan navigational behavior by directly mapping onto the representation (the street which goes to the right on the map will go to the right on the ground). One should remember also that a large part of the content of maps obeys conventions and involves some amount of arbitrariness.

Let's consider an illustrative case of such arbitrary choices. For centuries, we have been accustomed to look at maps oriented in such a way that north corresponds to the top of the page on which the map is printed. Remarkably, the old maps of the city of Paris attest for systematic preference for a bird's-eye view where the River Seine is taken as a vertical axis of symmetry. In the map by Truschet and Hoyau (also called Plan de Bâle), which was designed in the sixteenth century, the city is drawn as one could see it from an elevated point from a distance, with the Seine flowing down to the bottom of the map. The same is true for a number of other maps, all through the seventeenth century, such as the maps designed by Gomboust or by Merian (Figure 2.2). A plausible explanation for this preferred orientation may be advanced. Since buildings are drawn as three-dimensional, this particular vertical orientation is the only one that makes it possible to display the façades of buildings, and notably the façades of churches, as mostly noble monuments whose façades are always facing west, a drawing choice which highlights their most remarkable visual aspect.

Jouvin de Rochefort, in 1676, was the author of the first map of Paris where the Seine crossed the cartographic space from right to left. According to experts, this option seems to have been motivated by a search for originality with respect to the dominant one. In the eighteenth century, Father Delagrive explicitly recommended north-oriented maps as satisfying one of the main criteria of advisable cartography.

FIGURE 2.2 The Merian map of Paris (1615)

Cartographic truth and intelligibility

We know that there are cases where maps include distortions due to the designer's intention to create an erroneous representation in the mind of their users (see Monmonier, 1991). Apart from such tendentious cases, the quality of a map is judged essentially by its capacity to convey exact and reliable information, as would be attested by the actions performed by users on the spot on the basis of this information. A message communicated by a map is expected not only to be "veridical," but, still more significantly, to be "true."

A domain of cartography that reveals an interesting characteristic is the elaboration of maps intended to convey the vertical dimension of represented geographic objects. There is a repertoire of ingenious methods, such as those based on the use of contour lines to visually suggest differences of height by a gradient of colors, thus giving a realistic impression of relief. More interesting are the actual 3D maps that are built from a thermoformed plastic support, which materialize relief precisely by showing the elevation of every ground undulation, in particular rocky eminences and mountains. The designers of such maps commonly use two scales, one for representing distances on the ground (planimetric scale) and another for distances in elevation (altimetric scale). As a matter of fact, if the same scale was used for both representations, the heights would be visually less discernable. In order to make the map more striking, it is classic to have recourse to a different altimetric scale, which visually highlights the impression of relief (Figure 2.3). This is a remarkable situation. Whereas maps, due to their analog properties, are considered as representations which truthfully preserve the relations among distances, the use of

FIGURE 2.3 A relief map

a different altimetric scale introduces a distortion effect, which is in fact a kind of cheating, but conveys a more "veridical" representation of the third dimension. It is as if introducing some amount of falsity in the representational process helps the user to enjoy a more authentic cognitive experience.

The physical experience of moving in our environment is undoubtedly an irreplaceable way to collect information on surrounding space. Maps are also a source of spatial knowledge, especially for environments whose size renders them inexplorable, even by survey exploration. As a matter of fact, geographic spaces of special interest for humans are extended spaces, those spreading out beyond the pericorporeal space or the limits of the territories in which people usually move.

Geographic knowledge acquired from maps is generally considered as valid, based on the shared intuition of figural accuracy of the map relative to the represented territory. The analog properties of cartographic representations form the basis of their truthfulness. Nevertheless, our memory of these maps, once they are no longer available to our sight, is not protected from weaknesses, and our geographic knowledge based on maps may well prove defective, especially when we attempt to put together representations that have been stored in memory as independent pieces. A famous example of this phenomenon, as reported and analyzed by Barbara Tversky (1981), can be found in situations where people are invited to check for the spatial relationship between two cities, one in the northern hemisphere and the other in the southern one. For instance, which of these two cities – New York City and Santiago de Chile – is east of the other? With the greatest confidence, most people answer that New York City (on the east coast of North America) is east of Santiago (on the west coast of South America). In order to provide such answer, people perform some kind of mental alignment of North and South Americas and respond by relying on the presumed relation between the east coast (in the north) and the west coast (in the south). But they are wrong. Their judgment would be valid if the two Americas were perfectly aligned one to the other on the globe. In fact, South America is strongly shifted to the east relative to the North American continent, with the result that a good number of points of the west coast of South America are in fact to the east of the east coast of North America (Figure 2.4). Many other examples of such "distortions" of geographic memory have been reported, but these errors are due to the fact that fragments of geographic memory are penetrated and altered by rational knowledge, which unduly "correct" spatial data with which such knowledge does not match.

The usefulness of a map depends on the richness and quality of the information it contains, but paradoxically it is also dependent on the secondary information that the cartographer has decided *not* to include. The quality and value of a map imposes a selection among the large number of potential candidates to inclusion. The level of detail defined by the map designer is of course correlated to the scale used.

FIGURE 2.4 North and South Americas. Adapted from the article by B. Tversky (*Cognitive Psychology*, 1981), by permission of the author and Elsevier

Geographic knowledge relies on the idea of an optimal "grain" for the good reading and comprehension of the map. In general, the objective is to keep only the broad outlines and withdraw any earth undulation below a criterion value in order for the representation to be legible enough and allow users to create a useful mental representation of a geographic reality. The cartographer has thus the responsibility of *eliminating* pieces of information that are judged to be not relevant. Of course, the first information sacrificed is the real size of the territory to be described, which imposes onto the cartographer the need to decide the extent to which it will be shrunk by setting what will be the most meaningful scale to use. In this regard, one may refer with a smile to the cartographic ideal attained by the mythical map of a territory that would use a 1:1 scale. This fiction excited the imagination of many writers, from Lewis Carroll to Jose Luis Borges and Umberto Eco. This ideal, perfect, exhaustive map would only have one shortcoming, that is to be . . . unserviceable!

One may also smile when thinking of the marine chart, endowed with still more virtues, which inspires the praise of the sailors in Carroll's *The Hunting of the Snark*. The ideal chart in question is in fact just a plain blank sheet . . . The sailors are said to be pleased to have "a map that they could all understand." They add:

> What's the good of Mercator's North Poles and Equators, Tropics, Zones, and Meridian Lines? [. . .] They are merely conventional signs! Other maps have such shapes, with their islands and capes! But we've got our brave Captain to thank [. . .] that he's bought us the best – a perfect and absolute blank!
>
> *(Palsky, 1999)*

More serious are the concerns of today's cartography, which is committed to producing informative documents, but without necessarily being bound by the necessity of pure spatial analogy if the objective is to facilitate visualization and intelligibility of phenomena. This is the case for the cartographic transformations of positions, by which the variable that the cartographer aims at representing imposes distortions to a geographic plane. The surface is then decomposed into a set of elements on which the application of "forces" corresponding to the variable of interest entails expansions and contractions which reveal the phenomenon under consideration (Figure 2.5) (see Cauvin, Escobar, & Serradj, 2007–2008).

Today, there are expressions of concern about the fact that cartography might reflect a too traditional approach, fully dependent on Euclidean metrics and committed to the principle of homothetic reduction. Therefore, innovative efforts are being developed to offer new modes of representation, such as "cartograms," by which the map background is reconstructed by anamorphosis. The expected value of this type of geographic representation is to be more accessible to critical reading, in particular on the part of non-specialists. The method is described by its promoters as opening perspectives for a "participatory cartography" and more immediate access to the critical information contained in the map (see Lévy, 2004).

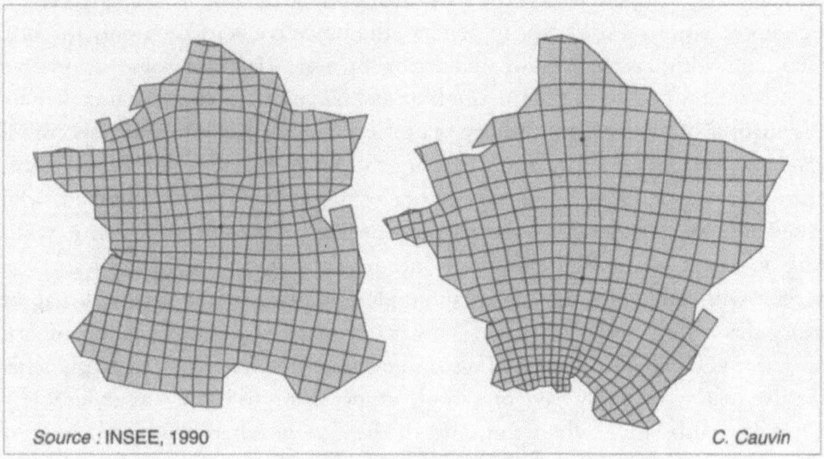

FIGURE 2.5 Cartographic transformation showing that the distribution of managerial staff in percentage in France (left) tends to be proportional to the total population (with slight overrepresentation in Paris), whereas the distribution of workers (right) shows a sharp contrast between the north and the south. Reproduced from the article by C. Cauvin (1998), by permission of the journal *M@ppemonde*

Very distant spaces

Quite different from geographic space are those spaces investigated by astronomers. By rising above the terrestrial globe and aiming at running across distant spaces, we discover that the third dimension imposes itself on a par with the other two, which are of primary importance in terrestrial geography. Moreover, we now cope with a concept of space that can no longer be figured out by the lay person. The mental effort to build a representation of this unaccustomed dimension of space has at least the value of introducing us to an intuition of the *infinite* character of space. We also approach a concept which adds to our cognitive dizziness – the concept of vacuum.

As in geography, we encounter the concept of distance, but the distances to be considered are of quite another order of magnitude. It is almost impossible for human understanding to create representations thereof. For instance, the distance from the Earth to the Sun is 150,000,000 kilometers, but its distance to the nearest star of the solar system – Proxima Centauri – is itself 270,000 times longer (which corresponds to four light-years). This reflects a dimensionality for which the human mind is not prepared. While geography and cognitive sciences are concerned by the relation between real and represented (cognitive) distances – a question that makes sense when one considers traveled spaces in our daily lives – the question becomes irrelevant when one deals with astronomical distances.

Stars have always been a subject of interest for humanity. Whereas the determination of their position on the celestial sphere has been a concern for astronomers since Antiquity, without any interest for the distances that separate them from the Earth, it is only in modern times that it was realized that stars are celestial objects located quite distantly from us. Constellations, as sets of especially bright stars arranged in more or less meaningful shapes, appear to be spread out on the celestial vault as if it were an even surface. Much time was needed for astronomy to make people understand that these shapes are only perceived as two-dimensional because of a projection effect. The stars that compose a constellation are separated from the Earth by quite different distances and constellations are in fact distributed across a three-dimensional space. To make the situation a bit more complicated, the groupings of stars into constellations are affected by the movement of the stars. The comparison between the sky map at the outset of the project "Carte du Ciel" at the end of nineteenth century and today's astronomical data attests for these changes.

In those vast spaces to which astronomy is dedicated, one finds familiar geometric concepts, such as the notions of ellipsis or elliptical trajectory, which are characteristic of planets' orbits according to Kepler's first law of planetary motion. In a dynamic perspective, a considerable challenge after Kepler was to account for the movement of celestial bodies on the basis of scientific observation. The forecast of such motion is at the core of the discipline devoted to the comprehension of celestial mechanics.

3

SPACE-RELATED PRACTICES

Scarce are the human activities that are totally unrelated to space. For some people, space is the very locus where a practice applies. This is the case of those occupations dedicated to the construction of buildings or the arrangement of environments in which human life takes place. For architects, urban planners, and landscapers, space is the "material" on which they operate and that they are invited to transform. This is also the case of people involved in cultural activities. From performing arts to the pictorial world, they develop their creativity in a spatialized world. This chapter reviews those spaces whose arrangement is subordinated to a creative intention, as well as those associated with the world of spirituality or beliefs. This will be extended to the sectors of literature which are concerned by the description of space.

Architectural and urban spaces

It is the remit of architects and urban planners to create and develop spaces for people's lives, work, and housing. Their productions can benefit large numbers of users, but they are sometimes planned for more specific populations, as is the case for environments designed for people with low vision or limited locomotor capacities.

Architects, urban planners, and landscapers are dedicated to space, its development, and the construction of buildings that fill it. Space is not only something to be contemplated, but a place for action and interaction for the humans who live in it. In this respect, it influences human action and determines the humans who inhabit it. This makes it critical for architects and urban planners to take into account the *use* that will be made of this environment. A critical concept in this respect is the concept of "legibility." This perspective guided Kevin Lynch (1960), an architect and urban planner who argued in favor of the legibility of maps, but also of the city itself, as its inhabitants experience it and build of it a coherent mental schema. The legibility of an environment enhances the level of empathy and comfort perceived by its users. It obviously has an impact on how people navigate through it and memorize it (see Montello, 2014).

The concept of space as "protagonist of architecture" has been illustrated in the book by Bruno Zevi (1957). An architect draws blueprints, which are nothing more than abstract projections on a horizontal plane of a combined set of walls. The blueprint describes a reality that nobody will actually see as such. But, as pointed out by Zevi, "architecture [. . .] does not consist in the sum of the width, length and height of the structural elements which enclose space, but in the void itself, the enclosed space in which man lives and moves" (pp. 22–23). And he adds: "Internal space, that space which [. . .] cannot be completely represented in any form, which can be grasped and felt only through direct experience, is the protagonist of architecture" (p. 23).

Architectural space is thus highly specific and cannot be defined in the same terms as pictorial space. A sensible architecture must take "internal space" into account. Furthermore, while the experience of architectural space takes place for people when they are in the edifices built by architects, it is the same experience that spreads out beyond, in the urban environment, streets, squares, parks, and gardens. By this way, "every building functions in the creation of two kinds of space: its internal space, completely defined by the building itself, and its external or urban space, defined by that building and the others around it" (Zevi, 1957, p. 30). The notion of continuity between both spaces is an innovative one. It is driven by a conception of space in which the human occupies a central place, in particular the human moving through built structures.

The concept of "inhabited space" is of increasing importance for architects, and more generally all people whose occupation is dedicated to the conception and planning of space. The idea of a "science of inhabited space" is well illustrated today (see Bonnin, 2007). It is articulated with the concept of "experienced space," as promoted by geographers. It includes the notion of spatial devices articulated with the representations and symbolic systems specific to the cultural environment in which they are designed.

Of course, humanity has not waited until modern times to practice architecture in close connection with people's representations, expectations, and beliefs. Anthropology provides a number of examples of architectural construction reflecting the mastering of meaningful spaces whose structure is intended to regulate human action and thought. In the spatial orientation and layout of constructions that form the living environment of a human group, it is well documented that the spatial design of living places is anchored in the group's myths and cosmogony that inspired them (see Karadimas, 2005).

Returning to the concerns of today's architects and urban planners, some of them are particularly dedicated to the conception and arrangement of large-size spaces, open to large numbers of users, typically identified as "public spaces." These areas are at the core of social life. They call for analyses intended to understand what kind of "uses" are attached to them. Most are urban spaces that people move

FIGURE 3.1 Misleading signage. (Photo credit: G. Cotteret)

through, and must be designed to optimize flows. Some spaces are to be thought out in order to accommodate large-size groups, and their layout must be arranged in order to secure inward and outward access. Spaces shared by large groups of people are those where signage is of particular importance. We all know by experience that uncontrolled directional signs may create disarray and misguidance in unfamiliar environments (Figure 3.1). Experts in signage are among the most useful "practitioners" of spatial matters (Mollerup, 2013; see also Denis, 2018).

As far as whole cities are concerned, the urban planners' challenge is to design neighborhoods and envisage how human activities are to be distributed by taking into account people's mobility. An important question for designers of an environment is that of the attractivity of various districts, in particular downtown vs. peripheral districts. Urban planning sometimes favors simplistic models which result in physical separation of activities in urban space, for example the separation between commercial and residential areas, with, as a result, the gradual relocation of commercial activities at the periphery, both cause and consequence of car use by urban dwellers. The development of these relocation processes are likely to affect urban morphology, and thus the perception and use of urban spaces.

Staging space

Traditionally, the stage of a theater offers a space delimited by a set where the characters' actions take place. From a purely geometric point of view, the stage is just a portion of a more extended and diverse space, which includes the theater hall and the whole building that houses the theater. The stage can be described by using the same units of measurement, in metric *continuity* with the space occupied by the audience. Nevertheless, it remains clearly separated from it from a functional point of view. In this well-demarcated area, particular events take place, distinct from those that happen in the hall. Theatrical space has its rites and codes, and even its spatial terminology, as is the case with the distinction between "courtyard side" and "garden side."

The history of theater is marked by practices that have greatly evolved, but has always recognized a particular status to the space in which the actors move. In the ancient Greek theater, the spectators gathered in the semicircular bleachers have in front of them a space structured into perfectly codified subspaces. Proscenium, that is the space occupying the whole width of a platform, is reserved for the actors, while the chorus is located in the orchestra, a distinct portion of space, circular in general, at the bottom of the bleachers. The chorus shares this space with the dancers, and an altar – the thymele – is located at the center. The two spaces have distinct functions, assigned to distinct players, those engaged in the act, and those who comment on it.

Although such differentiation is less marked, the stage of Italian-style theaters, as well as Elizabethan theaters, remains a well-demarcated space, preceded by a proscenium and encompassed by a set in trompe-l'œil supposed to create some illusion. Illusion, however, remains dependent on the willingness of the audience who is in general ready to accept it. In the history of theater, the sets were at certain times very realistic, whereas later they tended to become more sober and even minimal. But if geometric continuity between the actors' and the audience's spaces is real, scenography aims at fine-tuning the tricks intended to accentuate the cognitive and emotional separation between both spaces. Having their attention captured efficiently, the spectators lose the sense of being related to the stage by a continuous space. The space of the stage is then intensely experienced by the spectators as distinct from the space that they themselves are occupying. Mental immersion creates a feeling of "presence" (which can even be stronger still in the case of motion pictures and is one of the challenges of virtual reality devices).

Cinema is also an art of space. Scholarly analyses have been made of the processing of cinematographic space (see Agel, 1978; Rohmer, 2000). While theater offers a fixed window opened on a show (even if some devices make it possible to focus the audience's attention on a part of the stage area), cinema gives the possibility of changing the perspective on a filmed scene and varying the points of view imposed on the spectators. Furthermore, the physical space of the screen delineates an area, but the spectator is never disconnected from the narrative world even when it is temporarily off-camera.

This way, a film-maker (contrary to what happens on a theater stage) plays with an extended space while inviting the spectator to focus from time to time on subspaces where particular actions take place. This practice does not dismiss the persisting existence of the out-of-field parts of the world despite the fact that they are not in the audience's view, but instead they remain in the spectators' "cognitive field." A film director may successfully maintain the whole scenario of a film within a single space (for instance, a flat), and the excitement this causes will increase when the director manages to make the duration of the events exactly fit the duration of the story itself. One of the most striking examples is Alfred Hitchcock's *Rope* (1948). The spectator accesses the feeling of a veridical space, explored from all possible angles by a camera which adopts a continuously updated egocentric point of view, just as a person would do while visiting the actual flat.

Apart from such special cases, when a film invites the audience to get acquainted with several spaces involved in the same story, split into disconnected subspaces by the film editor, the spectator's challenge is to build a coherent "mental map" of the whole environment where the story takes place. Realizing the connections between separately displayed spaces may be a challenge for the spectator.

Museum space

A domain in which a kind of "staging" or sensible "setting" of objects in space is required is museography. The objective here is to find which layout may best highlight art objects or create conditions to make their meaning more evident. The architectural form of a museum itself is a parameter to consider. In this respect, it is interesting to realize how the spatial structure of museum buildings has evolved throughout the ages, from the "temple-like" model (such as the British Museum in London or the Altes Museum in Berlin) to more recent daring architectural forms, whether minimalist or monumental, but always based on the idea that a museum is itself an artistic expression of its time (the Milwaukee Art Museum or the Kunsthaus in Graz). Pioneering concepts have been launched, but not materialized eventually, such as the fascinating Museum of Unlimited Growth imagined by Le Corbusier in the early 1930s. As a spiral developing around a square schema, the building was planned to ensure a practically unlimited enlargement of the museum.

In a museum, the issue of spatiality is also critical in the conception of the rooms. Until the nineteenth century, a museum was conceived as a succession of rooms, treated as closed spaces, each of which being dedicated to a period, a style or a theme. The consequence for the visitor was the experience of a predetermined and thus highly constrained path. In more recent times, this conception was disused to the benefit of spatial distribution of objects offering the possibility of unconstrained journeys through the museum.[1]

In the rooms of a museum, a further spatial issue emerges, which is that of the positioning of the works relative to each other and relative to the background against which they stand out. In this respect, a principle has progressively imposed itself in the past century, that is, the concept of the "white cube," which provides the most neutral context for the exhibits. As vast galleries or more or less divided spaces, these environments offer their neutrality for the benefit of better showcasing the works. However, the spatial context in which a work is exhibited is never completely neutral. For instance, the Guggenheim Museum in New York, as a spiral developing against a helical ramp, is a concern for some artists for the hanging of rectangular paintings in an environment dominated by oblique rather than horizontal lines.

To break with the traditional hanging of works on the walls of a room, the idea has emerged to provide more visibility to the works by replacing each picture rail by a kind of socle or plinth, in such a way that paintings leave the impression that they are suspended above the ground and distributed in the three-dimensional space of the exhibition room. This approach is compatible with the principle of unconstrained paths for visitors. More and more, non-canonical museum environments are made available to the public. An example is the museum of the Centrale Montemartini in Rome, where industrial engines offer an unexpected context to marble statues of ancient Rome, resulting in convergences of interesting esthetic value.

Pictorial space and the issue of perspective

According to a certain conception of pictorial art, artists have a view of what should be the "focus of interest" of their paintings, and they develop adequate means to guide the spectator's sight to this focal point. To achieve this, painters will posit pictorial elements so that they favor this guiding process. The center of interest is generally characterized as the point of the painting towards which "lines of force" converge. These lines of force organize the pictorial work spatially and ensure its internal equilibrium. The horizon represented in the painting or the boundaries which underline the contrast between differently colored zones contribute to this organization, which is built mentally by the spectators as they examine the work.

How to represent the world space by inscribing it in the space of a painting? Before the emergence of abstraction, the artist's intent is to recreate as much as possible the appearance of the real. In particular, when the task is to represent depth, the challenge is to offer to the viewer a reasonable "illusion," while taking into account the two-dimensional structure of the canvas. The history of art shows that a variety of solutions have been found, reflecting a more or less marked commitment to obtaining this illusion. In Egyptian art, but also Chinese etchings or medieval art, the treatment of space is not overlooked, but its representation is not subordinated to the requirement of realism that will appear later as regards the rendering of depth. Lines are clear, flat tints of color prevail, shades are absent, frontal or lateral perspective is systematic, the world is represented in the same plane space, the scene is still. It is as if the painter or drawer was not guided by the intent of representing the world as it is available to an observer's eyes. Characters are usually juxtaposed and their sizes are determined by symbolic requirements (kings are represented taller than other people, whatever their position in the scene), without taking into account the proportions commanded by relative distances.

By the end of the Middle Ages, a more rational approach to the processing of space emerged, with an effort to locate characters in a frame that renders the architectural space veridically. The Renaissance then attempted to subordinate the representation of perspective to strict geometric principles. Painters and architects converged on an objective, that is, the perfection of divine laws should be expressed by the rules that govern geometry and the pictorial representation of the world. The rendering of perspective is then guided by a set of strict principles, in particular those operating in the creation of the depth effect, where the decrease of the objects' size is proportional to their distance to the front plane. Among the most significant postulates are those stating that the horizon is an absolute reference, that the projection

FIGURE 3.2 Vanishing point. (Credit: W. Gothe)

of a straight line is straight, that verticals remain vertical, whereas horizontals are vanishing. Parallel lines in the real space and not parallel to the plane of the picture must be represented by convergent lines, whose common point is the "vanishing point" (Figure 3.2).

While the representation of depth can be rendered by less mathematized procedures (as does, for instance, the background of *La Gioconda*), the primacy of perspective has guided numerous pictorial projects through the centuries. Two centuries after Brunelleschi and Masaccio in Florence and after the daring distortions applied to space by Uccello in *The Deluge*, Vermeer's paintings reflected the strong commitment of the Dutch Master to the linear perspective, to ensure rendering of the represented space in the silent frame where the painter and his model stand, as in *The Art of Painting*. The depth effects are dependent on a geometric construction that relies on rigorous combination of vanishing lines. Remarkably, through a rather limited production, Vermeer demonstrates his attachment to a small number of very similar spaces. The perspective applied to his pictorial arrangements always results in a highly expressive structure. Some authors have speculated that the perfection of lines was due to the use of a *camera obscura* by Vermeer, although this hypothesis is still discussed today (see Gutruf & Stachel, 2010; Steadman, 2001). A mostly valuable attempt has been made to use computing methods to reconstruct the real indoor space in front of which Vermeer's easel was placed.

Erwin Panofsky developed a thorough analysis of linear perspective, from the point of view of the history of art, in a text originally published in German in 1927 and later available in an English version (1991). For Panofsky, perspective is a mode of representation of the real which makes space visible according to the principles of Euclidean geometry. Perspective appeared at the time of the Renaissance as an attempt to make spectators access the equivalent of a direct experience of space. However, this stylistic tool remains a convention, associated with Western culture and conception of space, that is, an infinite and homogeneous continuum, which does not correspond to perceptual space, but to an essentially abstract reality.

In Chapter 1, we already noted the strong concern of Merleau-Ponty about space. Merleau-Ponty (1993) took up Panofsky's analyses, while rejecting his Cartesian and mathematized vision of space.[2] His endeavor was to understand what makes depth a special dimension of space. He argued that "four centuries after the 'solutions' of the Renaissance and three centuries after Descartes, depth is still new, and it insists on being sought, not 'once in a lifetime' but all through life" (p. 140). It is this analysis that drove Merleau-Ponty to the most sensible pages he devoted to Cézanne, a painter who "was seeking depth all his life" (as Giacometti pointed out), not by consenting to a Cartesian absolute space, but by approaching as much as possible perspective as it is experienced in natural perception. It is with this in mind that Cézanne could represent close objects smaller and remote ones bigger, in a process that mimics the natural mechanisms of human perception.

The issue of space was not forgotten, but was addressed in a totally new manner by Cubism. Cubist painters were attracted by the idea of superimposing the multiple aspects of an object, as captured from different angles of view. By doing this, they introduced temporality by figuring out the various facets of a reality captured while it is moving (or while the painter is moving around it). Abstract art was also faced with the issue of space, but in a new context, where space of the canvas supports spatial forms designed by the artist. One of the most impressive examples is found in the juxtaposition of colored surfaces by Mondrian.

Labyrinths and mazes

It is impossible to approach the concept of space without devoting some attention to labyrinths and mazes, those weird spatial objects intentionally designed to make their exploration a troublesome and possibly distressing experience.

Labyrinths and mazes are human-generated artifacts, resulting in spatial environments so complex that moving through them is a trial. They are intended to elicit particular spatial experiences, which sometimes impose ultimate challenges to people who accept running the risk. Nature offers environments that engender maze-like experiences, such as forests or deserts, or any unstructured environment, devoid of distinct landmarks, generating a feeling of enmeshment among useless cues. The design of those spatial artifacts recourses sometimes to sophisticated geometric rules.

Umberto Eco has been one of the experts in this domain, as attested by his literary works (*The Name of the Rose*, 1983) and essays (*From the Tree to the Labyrinth*, 2014). He proposed a classification of those special and intriguing spatial devices. The first one to be considered is the unicursal labyrinth, as illustrated in Greek mythology and in Gothic cathedrals (the most remarkable example being in Chartres; Figure 3.3). Here, the path is generally tortuous, but it does not include any branching or dead ends. Walking along its curves does not engender any feeling of uncertainty. It only calls for effort and persistence and is generally experienced as the symbol of spiritual progress. Distinct from the labyrinth, a maze (also identified as "Mannerist labyrinth" by Eco) is made of a number of paths, of which only one terminates at an exit while all the others finish as impasses. Moving along the tortuous paths of

FIGURE 3.3 The labyrinth of the cathedral of Chartres

a maze is a challenge because of the many ramifications leading to dead ends. Eco also considers a further case, which is a maze (also called "Hermetic labyrinth") made of an infinite network of paths, connected to each other by several nodes, with an evolving structure, and requiring repeated generation of hypotheses and bets. This is for sure the type of maze that imposes the biggest challenge to the person who ventures into it.

For Eco, from a cognitive and epistemological point of view, the labyrinth must be understood as "a paradigm for the organization of knowledge." In a metaphorical perspective, it illustrates a course aiming at reaching some form of knowledge. It is an allegory of "a multiple and evasive reality which escapes any unequivocal and all-encompassing structuration" (Lovito, 2013, p. 356).

Contrary to a first impression, a classic multicursal labyrinth (or maze) is not a hopeless situation. Just as there are more or less sophisticated algorithms to design labyrinths, there exist also resolution algorithms, some of which prove to be highly reliable. A practical method frequently recommended is to systematically follow the wall of the labyrinth by keeping one's hand (either the right or the left one, but always the same) in permanent contact with the wall. The exit of the labyrinth will be reached mandatorily, the only inconvenience being that this method will not necessarily result in the shortest path.

By the end of the nineteenth century, the French mathematician Gaston Tarry attached his name to the formulation of solutions for finding the exit of a maze. In an article with an impressive title ("Number of distinct ways of traveling in a single course along all the paths of a maze by passing only once by each path"), Tarry (1886) posited and demonstrated the "theorem of dead ends." He later proposed a solution based on an algorithm that one would characterize today as a depth-first search (Tarry, 1895).

Anchored in both geometry and metaphysics, the issue of labyrinths and mazes touches anthropological and social issues in which myths and beliefs are implied.

The mystical nature of labyrinths and mazes and the possibility that they offer accomplishment of a spiritual or initiatic journey are found in almost all cultures (see Conty, 2002).

Sacred spaces

In most religious traditions of humankind, from Egypt to Greece and through many far-off mythologies, a strong distinction is maintained between two spatial worlds. On the one hand, it is acknowledged that there exists a terrestrial space, created for living beings and inhabited by them. On the other hand, clearly separated from that space, there also exists an invisible world devoted to divinities and inaccessible to the senses of human beings.

It may also happen that in the space peopled by humans, some singular events occur, such as those by which gods manifest themselves and confer an exceptional property – a *sacred* dimension – to the part of the terrestrial world where this apparition took place. For humans, this divine incursion creates the conditions of a specific experience, the experience of sacredness, which tends to get structured in time and space according to codes and rituals shared by a community. To the concept of sacredness is attached the notion of an invisible greatness, which imposes itself on humans.

While ordinary people perceive space as homogeneous and distributed in an even and neutral way around them, religious people have a sense of sacred space as qualitatively different from the rest of the environment. A profane space may become sacred through processes like the appearance of a divine entity (this is theophany) or the disclosure of a sacred reality (this is hierophany). The place then becomes "marked." Within this territory, some zones may bear the print of sacredness in a more intense fashion. Codified restrictions may then be placed on its access. In his analysis of myths and religions, Mircea Eliade (1959) summarized this state of affairs as follows:

> There is [. . .] a sacred space, and hence a strong, significant space; there are other spaces that are not sacred and so are without structure or consistency, amorphous. Nor is this all. For religious man, this spatial nonhomogeneity finds expression in the experience of an opposition between space that is sacred – the only *real* and *really* existing space – and all other space, the formless expanse surrounding it.
>
> *(p. 20)*

Since very ancient times, the divinities are honored in places especially dedicated to worship. Although any place can be charged with this function, some natural places (springs, promontories, forests, caves) seem to inspire more than others the sense of sacredness. In particular, caves, dark places that are not adequate for welcoming humans' daily life and to which feelings of worry, strangeness and mystery are attached, have been amply described by anthropologists for their capacity to stimulate imagination and encourage "imaginary geographies." For Montello and Moyes (2012), the darkness that surrounds the visitor of a cave offers a kind of "transcendental affordance" which has no equivalent in any other environment.

In ancient Greece, by the end of the Geometric period (900–750 BC), the delimitation of sacred places dedicated to rituals was an activity of primary significance. The "temenos" (literally, "what has been cut off") designates the piece of land dedicated to gods. It is identified by its separation from the non-sacred land. Later, during the Archaic period (750–480 BC), religious thought turned towards more subtle and uncanny aspects of reality, which can hardly be seen or reached, such as ether, air, and marine abysses. The unfathomable and unreachable character of these spaces challenges imagination. It contributed to forging the Greek representation of cosmos, "a world set orderly and submitted as much to physical as to divine laws" (Cursaru, 2017).

But the most significant evolution during the centuries when Greece moved towards its Classical period was the transition from a system of thought where human activity is fully inscribed in a network of religious symbols to a system where reason begins to take precedence over myths. This evolution has been retraced by Jean-Pierre Vernant (1965) in his writings of "historical psychology" dedicated to Greek thought. In particular, the evolution pertains to space. Whereas space was traditionally organized by religious thought and related social practices, a new trend appeared in political life and uses, that is, the construction of a new space, "homogeneous, reversible, geometric," opposed to the space inspired by religion, "qualitative, differentiated, hierarchized." In its archaic representation, "space requests a center, a fixed point, having a privileged status, from which it is possible to draw qualitatively different directions" (p. 101). To this view of space is substituted a conception in which "instead of spatial values attached to a stationary center closed in on itself, one discovers the opposite values of open and movable space, designed for navigation, connections and transitions" (p. 140).

The upsurge of a rational representation of space was accompanied by an evolution of social practices, which was reflected in new modes of organization in the city, in particular a social space centered on the agora, where all citizens could perceive themselves as equal. "Behind Greek geometry and its homogeneous space based on symmetrical and reversible relationships, what is implemented is a new political practice and system of thought" (Detienne, 1966, p. 129).

Space of poetry and literature

Space speaks to all people who are dedicated to the creation of forms, and this includes creative people committed to the worlds of literature and poetry.

Midway between poetic adventure and philosophical journey, we come across the splendid book by Gaston Bachelard, *The Poetics of Space* (1964). Instead of "spaces" properly speaking, Bachelard delves into a set of "places" with a high emotional load, which contribute to both phenomenology and poetics of "the quite simple images of *felicitous space*" (p. xxxv). He first analyzes "the intimate values of inside space" (p. 3), such as the house, "our first universe," "the human being's first world." Other images of inside spaces are examined, such as drawers, chests, caskets, nests, shells, corners. By contrast, "immensity" is also discussed, as a "philosophical category of daydream [which] transports the dreamer outside the immediate world to a world that bears the mark of infinity" (p. 183). Bachelard also evokes the relationships between inside and outside, which form "a dialectic of division [which] has the sharpness of the dialectics of *yes* and *no*" (p. 211).

Bachelard refers to the dizziness that may sometimes be experienced in space, in the form of claustrophobia or agoraphobia. He aptly quotes the statement by Supervielle that "too much space smothers us much more than if there were not enough." This concludes Bachelard's fine attempt to establish a "topoanalysis," which would be "the systematic psychological study of the sites of our intimate lives" (p. 8).

Literature offers a number of variations on the issue of space, in particular immense, elusive, unfriendly space and the anxiety that it generates, as attested in a converging manner by two famous novels reported by literary criticism as originating from a similar inspiration. This is the case of the books by Dino Buzzati (*The Tartar Steppe*, 1952) and Julien Gracq (*The Opposing Shore*, 1986). In both novels, a distant desert area offers a context for a never-ending expectation. Time is at a standstill. The environment is not really described, but is suggested, hauntingly pervading the characters' (and the readers') minds, and achieving a sort of confusion between real and imaginary spaces. It is worth noting that before his career as a writer, Julien Gracq had studied and also taught geography at the University of Caen. An interesting attempt has been made to design a cartographic representation of the environments where *The Opposing Shore* takes place (Lacoste, 1987).

Quite different is the task of writers whose objective is to properly *describe* spaces that they expect us to learn about. We will mention later the case of Gustave Flaubert and the pains he took to achieve the most precise description of a geographic site (Chapter 11). But let's presently consider two splendid examples of a writing process guided by the obsession of accurate spatial description. The first case is that of H. P. Lovecraft (1976), not only the author of the extraordinary fantastical writings that made him famous, but also of a number of essays, including the mostly intriguing text providing an incredibly detailed description of Quebec City.[3] Quite excited by the setting of the city and stunned by "its scent of fresh tranquillity, antiquity and eternity," Lovecraft retraces his walking paths and offers a thorough guidebook for the use of other visitors. After a long historical introduction, he makes us walk along each street, reporting every detail of the topography and the appearance of each façade. For instance:

> At Rue des Sœurs we make a jog to the right into the narrow and picturesquely ancient Rue Notre-Dame, which shortly spreads out into a small quiet square lined by ancient houses, having a circular grass plot and fountain in the centre, and bounded on the farther side by an old steep-roofed chapel with graceful, needle-like spire. This is the original market place of the lower town.
>
> *(pp. 262–263)*

This detailed travelogue covers one hundred pages or so. Through this quasi-obsessional effort to describe precisely each street and each square, Lovecraft nevertheless experiences the limits of this literary challenge and sometimes completes his descriptions with sketches, as when he tries to describe the precise shape of the saddle roofs of the houses of Rue Saint-Jean.

The same obsessional effort to describe space is found under Georges Perec's pen. The writer introduces his book *Species of Spaces and Other Essays* (1997) as "the diary of a user of space." He reviews several "figures" of space, such as borders, inhabitable places, uninhabitable places, places where one has slept. For Perec,

> spaces have multiplied, been broken up and have diversified. There are spaces today of every kind and every size, for every use and every function. To live is to pass from one space to another, while doing your very best not to bump yourself.
>
> *(p. 6)*

Probably his most striking reflection is about how sight looks over space, while giving us a sense of relief and distance:

> When nothing arrests our gaze, it carries a very long way. But if it meets with nothing, it sees only what it meets. Space is what arrests our gaze, what our sight stumbles over [. . .]. There's nothing ectoplasmic about space; it has edges, it doesn't go off in all directions, it does all that needs to be done for railway lines to meet well short of infinity.
>
> *(p. 81)*

Perec developed an ambitious project for the description of several Parisian places. He minutely described the method that he himself used and that he recommended for future "describers." The exercises of observation and writing that he imposed on himself are reported as "practical exercises" in the chapter of *Species of Spaces* devoted to the street. They are conceived to invite a writer to describe the banality of the city including in spaces considered as trivial or insignificant. "Make an effort to exhaust the subject, even if that seems grotesque, or pointless, or stupid. [. . .] Force yourself to see more flatly" (pp. 50–51).

The descriptive method was illustrated by Perec in his book *An Attempt at Exhausting a Place in Paris* (2010), whose title attests to the stoical effort made by the author in this literary endeavor. Perec develops two modes of description. The first one is applied in the case of a stationary position, from a precise place, with the writer shifting gaze with a panoramic movement that typically embraces a crossroads or a square. When the description of the place is exhausted, Perec introduces notes on the events related to traffic or pedestrians' comings and goings. He used this writing method for his descriptions of the Mabillon crossroads and Place Saint-Sulpice, as observed from a café terrace.

The second mode of description is implemented when the observer is walking. The transcript of information is made during the progression and is particularly suited to the description of streets. In his initial program, Perec had committed himself to describe each of the places selected for this exercise twice a year for a total of twelve years. The most famous descriptions written with this method are those of Rue Vilin (where Perec lived during his early childhood) and Rue de l'Assomption (where he spent part of his youth). This close coordination between a space, the gaze that explores it, and the writing gesture intended to record a minute, almost impersonal trace constitutes a form of literary asceticism of which there are few equivalents.

Notes

1 The arrangement of museum spaces is further discussed in Chapter 13, this volume.
2 *L'Œil et l'Esprit* was the last book completed by Merleau-Ponty before his death. It was published in 1964. An English version is available in the collection of Merleau-Ponty's essays on esthetics edited by G. A. Johnson in 1993.
3 *A Description of the Town of Quebec* was written by Lovecraft in 1931. The text is one of the seventeen essays of Lovecraft assembled and edited by L. S. de Camp in 1976, entitled *To Quebec and the Stars*.

PART II

Spatial behavior and spatial representations

We now address space as a natural context for human behavior. Space, as a matter of fact, is a *place for action* (see Gärling & Golledge, 1989). Here, we no longer focus on those people whose endeavor or occupation is to manage, fit out, or equip space. Our object of interest is just each of us, as spontaneous "users" of space in everyday life, as we experience it through motion, exploration, or reasoning. Psychological science attempts at catching people while their spatial behavior is being implemented, with the postulate that behavior is related to internal representations that people have constructed of their environment. Not only is behavior supposed to rely on representations that preexist it, but it is thought to continuously generate novel representations in turn. Behavior is then conceived of as an essential ingredient in the formation of spatial knowledge, and spatial knowledge is seen as the source of enrichment of people's cognitive repertoire. Such permanent interfacing between behavior and representations is a requisite for implementing the functions that serve spatial cognition.

An approach aimed at understanding the mechanisms that underlie spatial behavior first requires sorting out the various phenomena subsumed by the concept of spatial cognition. A preliminary question consists of making relevant distinctions as a function of the various "sizes" of spaces that people happen to cope with. Of course, it is not the same experience to inspect small-size spaces, such as those that participants explore in table-top experiments, and geographic spaces calling for locomotion. They must be distinguished, while also taking into account the common features, if any, of their cognitive processing. The following chapters will also bring us to the main classifying concepts, such as the spatial frames of reference that help people to figure out and reason about relative positions of objects in an environment. Cognitive map is another central concept, which refers to the informational package built up during learning episodes in a spatial environment and maintained in the mind afterwards. Lastly, we will review the measurements that people apply to space, whether in terms of locations, distances, angles, directions, and the processes by which they update spatial information while traversing an environment.

4
CLASSIFICATIONS

Space can be intuitively broken down into subspaces of more or less extended sizes, which justifies not treating under the same heading behavior that takes place in restricted spaces and does not necessarily require motion, and behavior unfolding in wide spaces, such as geographic territories and environments that can only be explored by locomotion. Admittedly, there are particular cases where the absolute size of a spatial object has no reason to be considered, that is, when its properties are not affected. For instance, an isosceles triangle is defined by a set of properties, and these properties are not affected by the size at which this geometric object is designed. As a consequence, when someone has to reason about the relationships between angles or segments of that figure, the validity of reasoning is independent of the absolute size of the figure. This being stated, beyond the processing of such problems calling for formal reasoning, the psychological functions develop in a context where different sizes of space imply qualitative differences. The idea was introduced that there are distinct approaches to spatial cognition, depending on the order of magnitude of the particular space under consideration (see Montello, 1993).

Then, how to cope with the variety of behavioral expressions of spatial knowledge? Recognizing a place or a landmark, computing the direction in which a building stands, returning to the starting point of a route, finding one's way in an unfamiliar environment, finding a shortcut, estimating the distance to a given place, describing the best way to reach a monument to a foreign visitor, are only a few among the various tasks that solicit our spatial knowledge. The recognition of this variety justifies researchers' special endeavor to draw rational principles for the classification of spatial behavior.

Spaces (plural)

An elementary classification that comes to mind consists of grossly differentiating between "near" (or "proximate") and "far" (or "distant") space; in other words, spaces pertaining to what is designated by the adverbs "here" and "there," respectively. This is a way to contrast, on the one hand, some kind of restrained space, wholly available to our senses, and on the other hand, all remaining portions of space, which lay beyond a close perimeter and require motion to be reached and experienced. The first space is more likely to be perceived from a single angle, the angle of our sight, while the second involves a plurality of points of view, implemented during the exploratory behavior that it elicits.

However, the strong inclination of the human mind for dichotomies does not long resist this crude distinction and the need to introduce some nuances soon appears. It is not surprising that the first attempt in this direction has proposed considering at least one intermediary level between the close and the distant. In an illuminating chapter, Dan Montello (1993), a scientist from Santa Barbara at the interface of geography and psychology, recalls some of these classifications and proposes his own segmentation. He underlines that the reference to "small-" or "large-size" spaces implicitly pertains to a dimensionality referring to the person, more precisely, the person's body and the associated actions. Based on these considerations, Montello's classification includes three levels, plus one more, which has a particular status.

First of all, *figural space* is described as a space of restricted size, typically smaller than a person's body. It can be perceptually apprehended from the place where an individual stands without implying any move on the part of the person. In general, the objects within this space can be reached and manipulated. In experimental psychology, this type of space open to close perceptual access is illustrated by the situations where an experimenter leaves a participant in front of a set of objects displayed on a table. The participant is invited, for instance, to judge the relations between the objects, to modify their positions according to the experimenter's instructions, or more simply to memorize their locations. Objects represented on a computer screen are also considered to pertain to this type of space.

Then, Montello refers to *vista space* to describe a larger space, more extended than the human body, but that can be visually apprehended without locomotion. A room, a garden, and a square are examples of such spaces. The spatial relations between the objects contained in that space can be directly observed from the vantage point where an observer is standing.

An *environmental space* is largely more extended than the human body and can be said to "contain" it. Its extension is such that it can only be apprehended by locomotion and by collecting information that implies a plurality of points of view. It is the space of buildings, neighborhoods, and cities. Spatial relations within this space are not directly accessible, but are "built" from the exploration of the environment.

Although the distinction among these three spaces brings useful clarification, there is need to introduce a further class of space, which Montello labels *geographical space*. Like environmental space, this type of space is much larger than the human body. It cannot be apprehended by locomotion, but can only be known by the mediation of cartographic representations. It is the space of territories, countries, and continents. As a matter of fact, maps achieve some kind of "contraction" of large-size space to represent it in the form of figural space, accessible as a whole by visual perception. It is the function of maps to ensure figuration of environmental and geographical environments, which consequently become easier to explore and manipulate. But this figural space in which a map is inscribed has the capacity to offer to our perception not only large-size spaces (in a reduced form), but also, at the other extreme, microscopic spaces, unobservable to the naked eye, but displayable visually, such as the interior of a cell or the structure of a molecule. Without cultivating the paradox, we have no choice than to note that the acquisition of knowledge about the infinitely small and the infinitely large goes through the same symbolic representational medium (see Montello & Raubal, 2013).

If we consider Montello's spaces the most directly relevant to human navigation – vista and environmental spaces – specific questions will arise as to their articulation and the cognitive transition from one to the other. This transition is even more tricky since it implies the coordination between two frames of reference, the egocentric one, which is predominant in the case of vista space, and the allocentric one, which is associated with survey representation of environmental spaces (see Wolbers & Wiener, 2014).

A similar effort of classification, with some nuances by comparison to Montello's classification, has been proposed by Barbara Tversky (2003, 2005), another outstanding figure of spatial cognition research at Stanford and Columbia. Reference is also made to the human body, as well as to the motor system, as the privileged mode of exploration of space. For Tversky, it is relevant to consider four spaces, each one characterized by its functional properties: space of the body, space around the body, space of navigation, and space of graphics (or "space of external representations").

Body segments are distributed in space, according to a set of determinate and invariant spatial relations. Each segment can be identified by its particular shape. The body, which has its own spatial extension, is engaged in actions deployed in outer space. What representation do people have of their own body? What is inscribed in a person's memory is not only the visual appearance of the different body parts (face, hand, foot, etc.), but also the relational functions that these parts entertain with the outer world (the face may smile, hands grasp, feet kick, etc.). The visual characteristics of body parts and their functional characteristics are closely linked together in semantic memory. Mental representations of the human body appear thus to be organized around the representation of their functional significance.

Around the body is a space filled with objects available to perception and that people can reach from the place they are currently occupying. The issue then is to establish whether all regions of pericorporeal space have the same properties and whether that space forms a homogeneous whole. In fact, space is structured around three dimensions to which extensions of the axes of the human body correspond: above/below, front/back, and left/right. When someone has memorized an environment filled with objects situated all around and then has to declare in which direction is a given object, it turns out that response times differ according to the body axis on which this object is located. The fastest identified directions are those of objects located above or below the person. The head/feet body axis, which is asymmetric, is correlated with another strongly asymmetric axis in the world, the axis of gravity. The front/back body axis is processed less rapidly than the preceding one. This axis is also asymmetric, by clearly differentiating the part of the world that is and the part that is not visible and manipulable. However, it is not correlated with an asymmetric axis of the world. Lastly, the left/right axis, which lacks salient asymmetries, is the axis for which processing times are the longest. In an elegant program of experiments, Franklin and Tversky (1990) were able to demonstrate the essentially inhomogeneous character of the reference frame that people use, as determined by the structure of the body and the related actions.

For Tversky, the space of navigation corresponds to the space that we inhabit and through which we move. Its size is such that it cannot be apprehended from a single sight. It is composite, made of many elements, although it is not perceived as an unorganized set. People tend to conceive it as an integrated whole. Although this space may be explored along its three dimensions when we move through it, it is frequently thought of as two-dimensional, as if it were seen from above. In other words, it is the space most likely to be handled and represented in the form of maps.

Lastly, as also illustrated by Montello, Tversky's model includes special consideration for graphically represented space. Maps use graphic space to represent geographic space. But graphic devices can also display non-visual and non-spatial realities, such as time or abstract relations. Far from being limited to figuring out literal meanings, the space of graphics has the potential of conveying metaphorical meanings.

According to Tversky, the processing of these various spaces calls for distinct cognitive and motor functions and most likely implies distinct brain structures. For instance, the tasks involving cognitive processing of small-size objects, like the classic mental rotation tasks, involve the parietal cortex (see Kesner & Creem-Regehr, 2013), whereas the acquisition of spatial knowledge of large-size environments relies on the hippocampus and the temporal cortex (see Redish & Ekstrom, 2013).

Unity of spatial cognition?

The presumption of dissociation between the cognitive competences needed for the treatment of different sorts of spaces emerged about twenty years ago among the group of scientists who developed the most significant amount of research on spatial cognition, by closely associating psychologists and geographers of the University of California at Santa Barbara. In a study published in 1995, Roberta Klatzky and her colleagues submitted sighted and blind people to both sets of tasks, which involved either visuo-spatial processing of small-size stimuli or locomotion through a spatial environment. Factorial analysis revealed that both groups of tasks corresponded to distinct factors, a result which strongly suggested the independence of the two types of tasks (see Klatzky, Golledge, Loomis, Cicinelli, & Pellegrino, 1995). Similar analyses conducted by other research groups resulted in similar findings (see Allen, Kirasic, Dobson, Long, & Beck, 1996).

The issue of correlation (or dissociation) between different types of spatial performance when people have to process contrasted spatial sizes was investigated still more thoroughly by Mary Hegarty, from Santa Barbara. First, in a study reviewing all works published on this issue in the past twenty years, she realized that only a very few of them displayed such correlations, which in fact were overall rather small (see Hegarty & Waller, 2005). In order to document the issue still more comprehensively, she developed with her colleagues a systematic large-size study in which more than 200 participants were submitted to the two groups of tasks, which involved a range of experimental situations (see Hegarty, Montello, Richardson, Ishikawa, & Lovelace, 2006).

The first group of tasks was composed of classic psychometric tests that provide measures of "small-scale" spatial aptitudes. The tests involved, for instance, visual inspection of embedded figures in order to recognize geometric forms, visualization of figural transformations calling for mental rotation, or recall of sequences of simple figures (arrows) shown in different orientations. The cognitive processing of these small-size visual objects was thought to involve the concept of "figural space," according to Montello's (1993) classification. The second group of tasks required navigation through "large-scale" spaces, which needed to integrate successive views taken along a route. Three conditions of such tasks in "environmental space" were considered: (a) spatial learning from locomotor navigation in a real environment; (b) spatial learning from navigation in a desktop virtual environment; and (c) spatial learning from viewing a videotaped environment. In all cases, participants were submitted to tasks such as providing direction estimates for landmarks encountered during learning, straight-line distance estimates for pairs of landmarks, or drawing a sketch map of the route traversed. These tasks thus involved memorization of "environmental space." The study thus provided the possibility of assessing the relationship between performances collected in two contrasted types of space. Correlational measures were expected to establish whether individual capacities serving visuo-spatial processing of restricted spaces are also involved in (real or simulated) navigation through large-size environments.

Multivariate analyses revealed that the aptitudes brought into play on small-size spaces are predictive of performance in tasks involving large-size environments, a finding that attests that cognitive processing at different scales of space calls for capacities at least partly common. The findings allow us to reject both the hypothesis of a unitary model of spatial capacities and a model of total dissociation. The most plausible model is that of partial overlap between the aptitudes required by the two modes of spatial processing. An independent study completed by Maria Kozhevnikov showed that response accuracy in pointing tasks ("Imagine that you are facing the university. In which direction is the airport?") is predictive of performance in navigational tasks in a complex spatial environment (see Kozhevnikov, Motes, Rasch, & Blajenkova, 2006).

An interesting piece of information from Hegarty's study is that according to the factorial analysis of the data collected for large-size spaces, distinct factors appear to account for learning of spatial environments based on visual media (videotape and virtual reality), on the one side, and on the other side, learning achieved from navigation in real environments. It also appears that spatial aptitudes involved in the processing of restrained spaces strongly predict learning of environments displayed just visually, whereas its predictive value is lower for learning based on direct locomotor experience.

Other aptitudes have no predictive value at all for the learning of extended spaces, namely, verbal aptitudes and abstract reasoning capacities. It is also interesting to consider participants' self-reported sense of direction, as based on their responses to descriptors like "I am very good at judging distances" or, to the opposite, "I very easily get lost in a new city." It turns out that participants' sense of direction is predictive of performance in large-size environments, and still more markedly in the case of actual navigation than when visual media are used.[1] Lastly, men's performance is higher than the performance of women in some spatial measures (but not all). The largest gender differences are found for the mental rotation task and measures of environmental learning (from both direct experience and visual media) (see also De Goede & Postma, 2015; Galea & Kimura, 1993). Let's note also that men report a better sense of direction than women.

The Hegarty study points to a critical notion: visuo-spatial capacities typically required for the processing of small-size materials do contribute to performance on spatial tasks in environments of larger extent. This relationship is plausible if one considers that capacities required by navigation call for mechanisms that are at work in tasks involving visuo-spatial figures. Mindful navigation requires coding of spatial information that comes to our senses. It also necessitates maintaining this information in memory in the form of a transient, but robust enough, representation. Lastly, it obliges travelers to achieve mental transformations of this representation and make spatial inferences. These are capacities among those typically involved in the processing of "figural" information. Such competences belong to the repertoire of processes engaged by working memory, more specifically visuo-spatial working memory (or "visuo-spatial sketchpad"; see Baddeley, 2007; Logie, 1995).

One way to confirm such an interpretation consists of collecting measures of navigational behavior after a learning phase during which a person has been required to execute a secondary (concurrent) task calling for the resources of the visuo-spatial component of working memory. Would navigational performance be impacted by this task? This is the issue investigated by the authors of two projects conducted in independent research centers, and which have in common being completed in rich urban environments. In both cases, the experiments took place in the center of medieval cities with narrow, tortuous streets, containing a number of landmarks and other details, and in which it is quite easy to get lost while navigating.

The city of Padua offered the ground for an elegant study in which the experimenters first invited participants to learn a route downtown for about fifteen minutes, escorted by a guide, and then to follow the same route without any assistance. During learning, the participants had to perform a concurrent task, either articulatory suppression (repeating continuously a sequence of meaningless syllables) or spatial tapping (clicking repeatedly on keys of a tapping board according to a predetermined spatial pattern). In a control condition, pedestrians did not have to execute any concurrent task during spatial learning. When the participants had to reproduce the learned route, the experimenters noted the number of directional errors as well as the number of pauses, as an indicator of the uncertainty experienced by the pedestrian at a given point of the route. The analysis of data showed that the control group made fewer errors than participants involved in the dual-task conditions, but there was no clear difference between the two kinds of concurrent tasks (see Garden, Cornoldi, & Logie, 2002).

A similar experiment was conducted in the medieval city of Tübingen, more precisely its virtual reality version, in which a particular route had to be learned by the participants. One of three types of secondary task had to be executed during route learning: verbal (perform a lexical-decision task), visual (compare mentally visual figures), spatial (indicate from which direction a sound was coming). The recall of the learned route was made with a joystick which allowed the participant to move in the virtual environment. The measures were the time to reach the end of the route, the traversed distance, the number of stops, and the number of directional errors. The results showed that the secondary tasks that had a significant effect on performance were the verbal and the spatial tasks. The impact of verbal activity very likely results from the fact that it interferes with the participants' spontaneous tendency to mentally encode direction changes ("left, next right, next left," etc.). The distinction between the visual and spatial components of the visuo-spatial sketchpad is illuminating, since it reveals the elective effect of the spatial component on spatial learning (see Meilinger, Knauff, & Bülthoff, 2008).

Wayfinding

In an effort to circumscribe the concept of "wayfinding," the late Reginald Golledge, a major figure at Santa Barbara who we will meet again in later chapters, referred to "the process of determining and following a *path* or *route* between an origin and a destination" (Golledge, 1999, p. 6). This definition covers in fact a variety of behaviors and contexts. Gary Allen (1999) distinguished at least three classes of tasks likely to be indexed under this term: "travel with the goal of reaching a familiar destination; exploratory travel with the goal of returning to a familiar point of origin; and travel with the goal of reaching a novel destination" (p. 48). A variety of cognitive mechanisms are mobilized at the service of such tasks. However, in spite of their diversity, tasks collated under the label "wayfinding" have in common involving processes of *representation*, *planning*, and *decision*, and they aim at reaching destinations much beyond the immediate horizon of a traveler when this person initiates the route (see Montello, 2005). All these components are to be distinguished from the purely *locomotor* component of navigation, which addresses a distinct processing level, related to the sensory and kinesthetic inputs collected along the route. Here, the control consists of avoiding obstacles, staying on course, taking uneven ground into account, and regulating pace when approaching particular objects. These components become of primary importance in the moves of non-sighted people. For sighted people in the majority of travel conditions, they are taken charge of by automatic processing that calls for moderate amounts of cognitive resources, while these resources are implemented to a large extent in travel planning and control.

A classification of wayfinding situations has been proposed by a trio of enterprising young scientists at the University of Freiburg, Germany (Wiener, Büchner, & Hölscher, 2009). The resulting taxonomy is for sure the most rational and exhaustive one currently available. It is guided by an analysis of the constraints imposed by the tasks (highly variable from task to task and thus calling for different cognitive processes). It takes also into account a dimension that is crucial, although frequently neglected, that is, people's prior available knowledge on the environment through which they are about to travel.

The first level to consider pertains to the external aids, depending on whether they are available or not to a navigator. In the case of *aided wayfinding*, the navigating agent is offered the possibility of referring to external representations provided by various media. These media include maps, directional signage, route directions, as well as navigational aids provided by hand-held computers or route planners. In general, urban environments display directional information intended to anticipate the needs of information that visitors are likely to face. In sum, an organizing authority (the City Planning Department) provides in advance pieces of information that will help a visitor to build a travel plan online. In a minimal version, the path will be materialized by a line drawn on the ground that the visitor will only have to follow passively. A good example is the Freedom Trail, a four-kilometer uninterrupted red line which allows tourists to travel across the historic district of the city of Boston without any uncertainty. The cognitive cost of following the line is very low and does not call for any sort of spatial reasoning.

In some cases, aided navigation requires more demanding cognitive processing, in particular when a person has to consult map information, either before or during travel. A difficulty here is the need to match cartographic information with the ground that the map describes. Coordinating the survey view provided by the map and the frontal view of the environment has a cognitive cost, imposed by simultaneous treatment of allocentric and egocentric perspectives on the same spatial entities.

The second case to be considered, in contrast with the previous one, is *unaided wayfinding* where no such devices as those mentioned above are available to the navigating person. Humans have many opportunities to experience this kind of navigation, but it seems to be the only mode available to other animal species. Several variants have to be considered (Figure 4.1). While lacking any external assistance, this kind of navigation is not aimless of course, but it is the nature of the goal that will differentiate between two subcategories, quite distinct from a cognitive point of view. The distinction here is between wayfinding guided by the aim to reach a goal located somewhere in space and wayfinding when the objective is not properly spatial, for instance when a person performs recreational walking without any specific destination.

In the latter case, which is a form of *undirected wayfinding*, a goal is not specified. If the environment is unknown to the moving agent, wayfinding will be limited

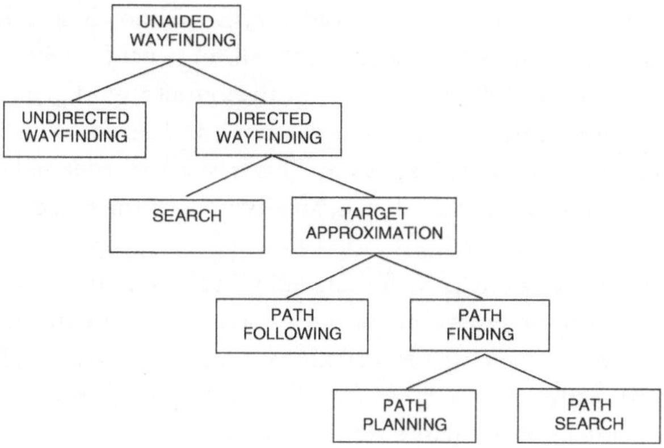

FIGURE 4.1 A taxonomy of unaided wayfinding tasks, as proposed by J. M. Wiener, S. J. Büchner, and C. Hölscher (2009)

to mere exploration, with the main objective of creating a representation of that environment, possibly still a crude one at the beginning of the wandering experience. But this form of wayfinding can also take place in a familiar environment. Examples are urban strolling without planning, guided online by encounters with spots of interest, or pure leisure walks in familiar countryside.

Contrastingly, *directed wayfinding* is motivated by a specific intent, that is, to reach a destination. But, again, a distinction must be made, depending on whether the navigator has any prior knowledge of the localization of the target. Either such knowledge is available (or can be inferred), or the person has no such knowledge and must engage in some kind of a search. The latter case will be examined first.

Directed wayfinding subordinated to *search* is illustrated by the cases where a moving person aims at reaching a specific goal. Two such cases can be considered. The first is when the navigator has some amount of knowledge of the environment in which a search will take place, but ignores the localization of the aimed target. The authors of the taxonomy give the example of a situation where you would be looking for a friend who is in a restaurant of a district you know very well. Being well informed about all the restaurants and their localizations, you are nevertheless still searching for your friend. You are familiar with the environment, but simply have no idea of where your target is located. In the second case, a similar search may be implemented, but in an environment of which you do not know the spatial structure. An example is that of a fireman who knows that a person is somewhere in a burning house. The fireman does search for a specific person, but has no idea about the person's location, nor the structure of the environment to explore.

In another case, directed wayfinding is motivated by the objective of reaching a target. In *target approximation*, the navigator has clear representation of the aimed target. But while the target is well identified, the navigator may face two distinct situations depending on whether he/she knows the path (or a path) leading to the destination. If the navigator knows this path and is able to retrieve it from memory, the task can be identified as *path following*, which only requires relating information captured in the environment and retrieved pieces of knowledge. In many such cases, the operation is automatic and does not call for special attentional resources, as for those routes that we follow routinely from home to workplace and *vice versa*. The other, more interesting, case is when the navigator has no prior knowledge of the route in memory and must make it up. This is a task of *path finding*. It is in this creative task, not free of risks, that the last subdivision of the taxonomy of Wiener and his colleagues is introduced.

This final distinction among path-finding tasks is related to the type of knowledge that a navigator has on the environment to be traversed. In a familiar environment in which the location of the target is well known, a person may simply be unaware of the path that directly leads to it because he/she has never had to reach it from the point where he/she now stands. This situation imposes performing *path planning*, a task that will force the person to retrieve in memory a representation of the environment and compute a novel route, which will require implementing cognitive resources (inference, reasoning, distance estimation, etc.). The situation will be even more difficult for a navigator who is in an unknown environment, and only approximately knows the target location, but without having any knowledge of the space to be traversed to reach it. Here, the navigator needs to perform a *path search*. Without any element to generate a reasoning process to plan a path, the person can only rely on heuristics to approach the target by successive iterations, in a potentially uncertain procedure. This is the situation of a tourist who is unfamiliar with the streets of Paris, but will be able to direct him/herself toward the Eiffel Tower by a succession of moves guided by the remote sight of the monument. This procedure is only possible when the target remains visible from a distance more or less continuously. If the target disappears temporarily from the view field, its position has to be recalculated by taking into account the moves executed in the meantime, before resuming progression toward the target once it is in sight again.

A navigational episode may well combine several tasks referenced in the taxonomy. Suppose that I have to join a friend in a hotel of a Parisian district where I never go. To reach the district is a task that pertains to *path planning*. Once I have attained the district, the task is likely to turn to a task of *search in an unknown environment*. Another example is that of a driver who parks his/her car by the entrance of a natural park, then starts strolling around the trails, which is to be considered as *undirected wayfinding in a known environment*. When the stroll comes to an end, the person will return to the parking lot by completing *directed wayfinding guided by target approximation*. The differences between the various tasks considered here have consequences on the moving agent's cognitive involvement.

Note

1 This point had already been evidenced by the same researchers in a previous study while they were developing the Santa Barbara Sense of Direction Scale (SBSOD) (Hegarty, Richardson, Montello, Lovelace, & Subbiah, 2002). Other similar scales were proposed by other researchers (Kato & Takeuchi, 2003; Münzer & Hölscher, 2011; Pazzaglia & Meneghetti, 2012; van der Ham, Kant, Postma, & Visser-Meily, 2013).

5

FRAMES OF REFERENCE AND COGNITIVE MAPS

Let's consider the case of a person equipped with every sensory means to perceive the surrounding world, standing in an environment filled with objects. How will this person build and organize knowledge of this environment? Beyond the visual appearance of the objects, what does the person ever know of their spatial distribution and relative positions?

Two relevant pieces of information are to be encoded in order to characterize an object's location. First, at what distance is the object located? Then, in which direction is it? These questions can only receive answers if relevant cognitive computations are executed relative to a *reference point* selected by the observer. The first candidate to such referential function is the observer him/herself or, more precisely, the point of space that he/she occupies. In this case, the direction in which the object can be found will be determined relative to the observer's body axis and the distance will be calculated from the place occupied by the observer. Another option is to use some other reference point, for instance a second object present in the environment. Angular and metric computations will take this second object as a reference. In both cases, the questions about distance or direction will be inscribed in a "reference frame." This notion – "egocentric" or "allocentric" frame – becomes crucial when the observer has to communicate with another person about the objects contained in that space. However, prior to any communicative act, the concept of reference frame imposes itself as a cognitive reality, a prerequired one for building new spatial knowledge. Substantial scientific knowledge has been established in the past twenty years, in particular about the brain encoding of the functions associated with these reference frames.

The existence of two kinds of spatial frames opens the issue of their acquisition and the chronology of their availability to the cognitive system. The scientist will then be facing the correlated issue of the variety of modes of acquisition of spatial knowledge and, beyond that, of its integration in long-term memory.

Special attention will be devoted to the concept of the "cognitive map" which has been used for seventy years or so to refer to the particular form under which people encode spatial information extracted from their environment and use it to guide their moves. People can also convey the content of this representation to others, either graphically or verbally. The metaphor has enjoyed great popularity in psychology, while requiring researchers to make its limits explicit. If this map is said to be "cognitive," it implicitly suggests that it is not material in nature, like printed maps in atlases. But if it is qualified as "map," it is because it validly expresses the spatial relations between the elements that it represents.

Frames of reference

Where is the bus station? Where is my car? The reference systems that will allow us to have a representation of these locations and find our way to them offer a choice between two candidates for this referential function, that is, either the person who is asking these questions, or a third object.

In the first case, the reference is said to be *egocentric*. My car is at a location that is defined relative to the position I am occupying presently, at the time I am asking the question. The egocentric frame of reference is, by nature, dependent on the transient position of an agent with respect to whom the question is examined.

Another case is when the agent takes an external fixed object as a point of reference. Any object can be used for this purpose. For instance, it may be a building as part of the environment where the agent stands. The position of the object considered in the questioning will be defined relative to the other object taken as a landmark. The issue is no longer to know where *my car* is relative to the place where *I* stand, but relative to *the fountain* of the square where I have parked my car. The question is then treated within an *allocentric* frame.

The analysis of the properties of the systems of reference has been undertaken in converging manners by the authors who explored this concept (Berthoz, 2000; Klatzky, 1998; McNamara, 2003; Montello, 2005; Paillard, 1991). Using one or the other system reflects a cognitive act that is guided by the same objective, that is, defining the position of an object relative to a reference point. The selection of this point makes the difference. For scientists, the issue then is to elucidate the cognitive competences brought into play and the brain structures that control spatial frames.

Although the present chapter focuses on the egocentric/allocentric duality, one should also mention the existence of a third system, which shares with the allocentric perspective the capacity of coding a location independently of the position of the observer. It does so without referring to any landmark present in the environment, but relative to a system of axes distributed over an extended space. The best example of such a system is provided by cardinal directions and coordinates defined in terms of longitude and latitude. Whereas landmark-based allocentric referentials are more frequently used in small-size environments (figural or vista spaces), coordinate systems are privileged tools to localize objects in large-size spaces.

When they are applied to the earth, the reference system is said to be *geocentric* (see Montello, 2005).

An example of experimental setting for the study of spatial reference systems is the following (see Committeri, Galati, Paradis, Pizzamiglio, Berthoz, & Le Bihan, 2004). Participants are presented with visual scenes on a screen where objects are distributed at various positions in a square area surrounded by a large three-wing building. While viewing those scenes, the participants are required to provide judgments of relative distances. In the egocentric conditions, they simply have to indicate which of two designated objects is closer (or farther) to themselves (as observers). Two allocentric conditions can be arranged. In one of them, distance judgments must be made relative to a third, distinct object also present in the area. In the other, participants have to judge distances of the objects to the building. The difference between the two conditions is that in the first case, the reference is an object likely to occupy different positions in the area, while in the second, the reference object is a permanent one, constituent of the structure of the environment.

The classic behavioral measures in this kind of experiment include error rate and response time. Would it be the case that decisions regarding distances are subjected to different cognitive constraints depending on whether they are made in reference to oneself, to another object, or to a building? The experiment by Committeri and her colleagues did not reveal any significant differences in this respect. In all three cases, the error rate was similarly low, and response times did not differ significantly, thus suggesting that different spatial frames involve comparable processing cost. Such an outcome invites researchers to explore further the issue and check whether the brain activity that accompanies these tasks is similar. The data here show that the cerebral activity underlying behavioral decisions reflects distinct patterns and suggest that processes resulting in similar behavioral responses may follow distinct neuronal paths. This is the subject of Chapter 7 in the next part.

Interactions

The egocentric perspective makes it possible to express the relation of oneself (as observer) to the objects of the environment. The allocentric perspective expresses the relation between objects (which serve as reference points to each other).

The same duality is found in other contrasts frequently called upon as regards spatial knowledge and navigation. The egocentric perspective is related to the notion of knowledge built sequentially, through locomotor experience which makes it possible to discover an environment gradually. The allocentric perspective, which is not dependent on a "first-person" exploration, is more akin to simultaneous knowledge collected from maps. Another contrast also related to those mentioned above, and which will be amply discussed in Part IV, opposes "route" and "survey" perspectives. These are two modes of apprehension of spatial realities, which offer two views of the same environment calling for distinct cognitive resources, and resulting in different verbal descriptions.

In the preceding paragraphs, we have only considered the conceptual differentiation between the two perspectives and their respective implementation in cognitive tasks. But what about the situations in which the two systems are brought into play in the course of the same task? Or in which a person has to switch from one perspective to the other?

In navigational tasks (where traversed spaces are frequently simulated by virtual reality devices for practical reasons), the possibility of changing from an egocentric frame to an allocentric one has been evidenced in the learning of moves, including moves in three-dimensional space (in conditions simulating weightlessness). The fact remains, however, that the optimal situation for memorizing a three-dimensional maze is the one where the egocentric vertical (of the person) is aligned on the allocentric vertical (of the environment). Of course, the updating of spatial knowledge is much easier when correspondence between the two reference frames is ensured (see Vidal, Amorim, & Berthoz, 2004).

An illuminating set of experiments on the relations between reference frames has been conducted by Kinga Iglói and her colleagues at the Collège de France, by examining the spontaneous use of egocentric and allocentric strategies within the same task, and the possibility of switching from a strategy to another (Iglói, Zaoui, Berthoz, & Rondi-Reig, 2009). The experimental situation consisted in navigating along alleys of a virtual star maze. During training trials, the departure point of a path was at the extremity of a branch of the maze and the path ended at the extremity of another branch, where a reward was delivered (Figure 5.1). The authors noted that the optimal path that would be eventually discovered (from alley 1 to alley 7), then stabilize in the course of learning, may result either from an egocentric strategy (by coding the sequence of body turns leading to the goal) or an allocentric one (by using environmental cues).

In order to identify which strategy was implemented during learning, the authors asked the participants to reach the reward place, but now starting from

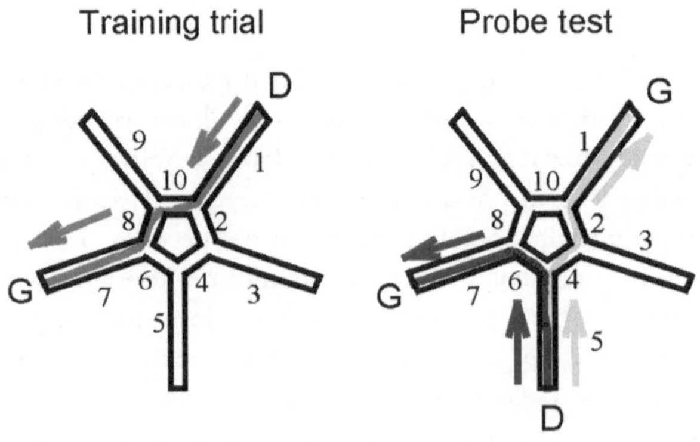

FIGURE 5.1 Paths in a virtual maze. D: Departure. G: Goal. Reprinted from the article by K. Iglói, M. Zaoui, A. Berthoz, and L. Rondi-Reig (*Hippocampus*, 2009), by permission of K. Iglói and John Wiley and Sons

alley 5. They were expected to see two kinds of responses emerging during test trials. Either the participants would reproduce the same sequence of body turns, which would bring them to alley 1, or they would follow a new trajectory which would bring them (correctly) to alley 7. The participants who adopted the first strategy (repeat the previously used sequence of turns) were considered as being guided by an egocentric strategy, while in the other case, where they appeared to rely on their knowledge of the global configuration, they might be considered as displaying an allocentric strategy. To be complete, a third class of responses also emerged, when the participants started with the egocentric strategy, then corrected the trajectory using environmental cues, thus demonstrating that they turned to an allocentric perspective. The most remarkable phenomenon was the shift from strategy to strategy along the trials of the probe test, which occurred in similar proportions in both directions.

In a further phase of the experiment, the participants were in conditions where the strategy was imposed on them. People here appeared to be able to switch to another strategy, in either direction, without impairment of their performance (reach the reward place). To summarize, there is every reason to believe that the participants acquired both egocentric and allocentric knowledge from the beginning of learning and were able to use them flexibly as a function of the conditions imposed later.

Imagined perspectives

There is no doubt that anyone immersed in a new spatial environment spontaneously develops an egocentric perspective on it. An allocentric perspective on the same environment, which results in that particular output termed a "cognitive map," calls for further, more demanding elaboration. Before looking at that specific kind of mental representation, there is a further aspect of egocentric perspective that still needs to be considered.

Each of us, with his/her own egocentric view of the world, interacts with other people who have their respective egocentric views of a shared environment. The question here is that of our capacity to adopt the perspective of other people. In order to view the world through the eyes of a person standing at another position than ours, the most straightforward solution consists in moving to that position and aligning oneself with the other person's perspective. In many situations, however, it is less costly to operate such decentration by just *mentally* moving to the place occupied by the other person and trying to imagine the world as seen by the other. Verbal communication about the shared representation is then possible, in particular if a collaborative task has to be executed by several agents in the same environment. Monitoring the linguistic outputs produced in such contexts is a challenge that we will discuss in Part IV. But upstream from language, it is in order to understand the representational mechanisms by which people take a new perspective on a scene, replacing an egocentric frame of reference by an "othercentric" one (see Duran, Dale, & Kreuz, 2011).

It is generally admitted that people, when needed, are able to change their points of view and adopt mentally novel spatial perspectives. Perspective-taking has a cognitive cost, but the idea is that by combining together spatial information taken from different viewpoints, people will enjoy the possibility of forming richer representations of their environments. They should therefore be able to achieve more accurate assessments of directions and distances to objects from yet unexperienced viewpoints.

One of the first attempts to explore this question is due to Amy Shelton and Tim McNamara (1997). Their participants were presented with a spatial display of seven objects spread out on the floor of a room. They were given the opportunity to learn the display from two orthogonal views (0 degrees and 90 degrees) (Figure 5.2). After learning, they were required to make judgments of relative direction using their memory of the layout. They were invited to imagine that they were standing near an object and facing another one, and were asked to point to a third object. For instance, "imagine you are standing at the jar and facing the shoe; point to the clock." Pointing responses were provided by participants on a computer screen by drawing a line indicating the direction to the target object (from the center of a circle representing the standing point, with the imagined heading pointing vertically up).

The difficulty of such a task results from the combined processing of two mental transformations over two angles. One is the angle between the orientation of the array and the perspective to be imagined. The other is the angle between the imagined perspective and the direction to the target object. The results clearly showed that judgments were faster and more accurate when the imagined heading was aligned with one of the studied views (for instance, "at the book, facing the wood" or "at the shoe, facing the pan") than when it was misaligned to either ("at the wood, facing the jar"). In terms of accuracy, angular error increased as pointing directions varied from in front to behind the participant. Front right and

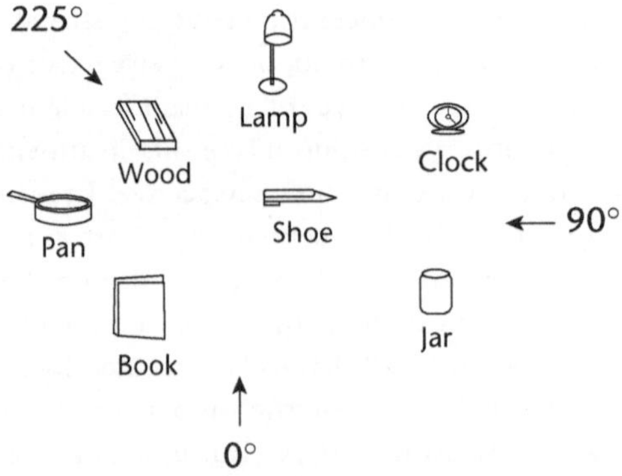

FIGURE 5.2 A spatial layout. Reprinted from the article by A. L. Shelton and T. P. McNamara (*Cognitive Psychology*, 2001), by permission of Elsevier

front left directions resulted in more accurate responses than behind right and behind left, as well as side directions, and sides were more accurate than back. Thus, imagining a perspective that has not yet been experienced has a cost, and this cost depends of the amount of discrepancy between the imagined perspective and the actual direction.

It is tempting to conclude that spatial memory will be enhanced when more than one perspective is taken on an environment. However, considering the data reported above, it is unlikely that just two perspectives would suffice to allow someone to build a fully consistent allocentric representation. According to the authors of the study, two egocentric views of a space do not generate a single viewpoint-independent representation in memory, but rather two viewpoint-dependent representations. There does not even seem to be any evidence that up to three stationary views of a spatial layout can lead to the formation of an orientation-independent mental representation. It also appears that the geometry of the room plays a role in how a spatial layout is mentally represented. The misalignment of the display with respect to the walls of the room (a factor likely to interfere with the global reference system) creates a context that makes the pointing task more difficult (see Shelton & McNamara, 2001).

Similar tasks intended to assess people's ability to imagine alternative perspectives were proposed by other authors. In particular, Kozhevnikov and Hegarty (2001) conducted experiments in which participants had to anticipate the appearance of a fixed array of objects after a change in their perspective. They used a configuration of seven objects drawn on a sheet of paper. The display was fixed to the wall directly in front of the participants (Figure 5.3 a). The participants were invited to imagine being at the position of one object (the station point) facing another object (defining the imagined heading or perspective within the array) and asked to indicate the direction to a third (target) object. For instance, "imagine you are standing near the stop sign facing the house; point to the traffic light." Pointing responses were collected through a device similar to that used by Shelton and McNamara (1997, 2001). A similar task used a map showing five geographic landmarks (Figure 5.3 b). For instance, "imagine that you are at the zoo facing the airport; point to the university."

Measures of pointing accuracy revealed that angular error increased with the angular deviation of the participants' imagined heading (perspective) from the orientation of the array (Figure 5.4). The number of errors increased significantly when the angular difference exceeded 90 degrees. The authors of the study noted that perspective-taking abilities are highly correlated with spatial visualization abilities (such as those involved in mental rotation). However, both cannot be considered as identical. Perspective-taking tasks and tasks calling for mental manipulation of visual objects are separate, and probably do not reflect the same construct.

Other experiments have examined perspective-taking capacities in people of various ages (younger, middle-aged, and older adults). The results showed that performance decreases across age groups, and confirmed that for all age groups, there is an increase of errors when perspective change exceeds 90 degrees (see Zancada-Menéndez, Sampedro-Piquero, Lopez, & McNamara, 2016).

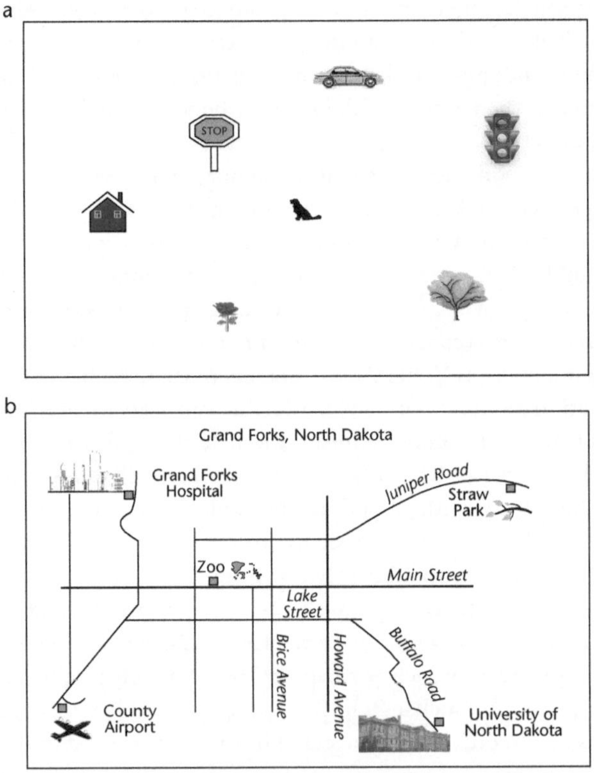

FIGURE 5.3 Two configurations. Reprinted from the article by M. Kozhevnikov and M. Hegarty (*Memory and Cognition*, 2001), by permission of M. Kozhevnikov and Springer

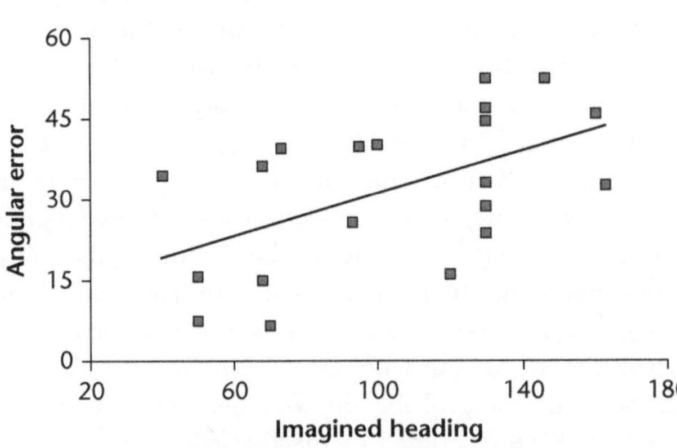

FIGURE 5.4 Absolute pointing error as a function of the angular deviation of imagined heading (perspective) from the orientation of the array. Reprinted from the article by M. Kozhevnikov and M. Hegarty (*Memory and Cognition*, 2001), by permission of M. Kozhevnikov and Springer

One should also mention a very special case where a person is inclined to take more than a single perspective on an environment. When a person is invited to learn a path through an unfamiliar environment, information is typically processed from an egocentric perspective. When the same person is informed that the return trip will follow the same path, a strategy may consist of making pauses at particular points, turning back by 180 degrees, and catching a glimpse of the appearance of the scene at this point. This is a way to collect an advance view of how the scene will appear in the return trip. Experimental demonstration of the benefit of that strategy (exposing oneself to an egocentric perspective known to be experienced later) is available. Compared to control participants, travelers who are invited to use this backward strategy show higher performance, whether it is measured in terms of the number of wrong turns at choice points, the distance traveled on off-route excursions, or the number of missed choice points. The first two measures are correlated with the angular error produced by the participants when they reach the end point and are required to point to the starting point. The results attest to the profit obtained from preparing oneself to the return trip by creating an anticipatory experience according to the egocentric perspective developed during that trip (see Heth, Cornell, & Flood, 2002).[1]

Cognitive maps

The expression "cognitive map" is used to refer to the particular form of mental representation that contains information on spatial relations among objects in an environment and provides people with the possibility of orienting themselves in that environment. The concept was created in 1948, under the pen of Edward Tolman, professor at Berkeley, in his in-depth review of learning mechanisms, in particular in rats as they are progressing down mazes. In his paper, Tolman engaged in direct confrontation with the proponents of the still influential theory that accounted for learning by the establishment of mere stimulus–response connections. He promoted the idea that during learning, "something like a field map of the environment gets established in the rat's brain" (p. 192). For Tolman, the animal is supposed to learn more than a succession of motor responses associated with a succession of external stimuli. Knowledge of the maze builds up from active exploration of the environment, without being simply dependent on reinforcement. The same idea underlies another notion that was born in Berkeley, the notion of "latent learning."

The cognitive map is one of the most appealing metaphors that has ever been created in the history of psychology, and one that achieved the greatest success. The metaphorical character of the concept is recognized by most researchers who adopted it, even though, as with any metaphor, it has its inherent limitations. A cognitive map is not equivalent to a page cut out from a geographical atlas or a city map. It does not include all their details. It contains portions less clear than others, or distorted. The cognitive map metaphor is, however, fully justified if one admits that the internal representation pointed to by this term has a *function* quite similar to the function served by physical maps in guiding human travels. In this respect, a cognitive map fulfills several of the essential functions of human cognition: to help recognize places, compute distances and directions, and assist people's wayfinding in an environment.

A cognitive map may include pieces of information that are not cartographic in nature. Information is not necessarily organized in the form of a coherent structure based on spatial coordinates. This internal representation is then likely to look less like a map than a *collage*. This view, which was promoted by Barbara Tversky (1993), offers the possibility of accounting for the fact that cognitive maps may not display the coherence and internal organization of map-like documents, but, like collages, they include figural information that may be only partial and combine several perspectives. The metaphor of "cognitive collage" might then prove to be more appropriate than that of cognitive map.

This approach does not justify any postulate of the intrinsic spatiality of cognitive maps, nor a spatiality of its neuronal medium. There is no indication of topological correspondence between the objects present in an environment and their brain representation. However, an important feature must be stressed. In a physical map, every point is characterized by its relation to any other point of the map. This confers a computational value that one also finds in cognitive maps (see Nadel, 2013). In order to be qualified as a cognitive map, a spatial mental representation must be endowed with a function that extends beyond mere guidance by reproduction of known routes, but it has to help a cognitive agent to define novel and suitable paths (like alternate routes, shortcuts, or detours).

Following this analysis, a cognitive map is to be considered as essentially pertaining to the allocentric reference system. It is independent of the particular viewpoint of an observer on the environment. It codes *configurational* information. In addition to conveying information on constituent objects of an environment and their appearance, it expresses their spatial relationships. It speaks of space, of the elements contained in it, and, still more importantly, of the relations among these elements.

In line with these conclusions, one would expect that when people retrieve from memory the cognitive map of an environment, the survey representation currently activated is independent of any particular orientation of these people in the environment. But does this exclude that the internal representation is entirely disconnected from the global reference frame of the environment? We should be aware that if knowledge of spatial relations among landmarks of an environment is acquired from locomotor experience, a good deal of such knowledge is also based on the processing of map information. According to the current practice in Western culture, maps display locations within a north-oriented reference frame. Could then it be the case that a person physically facing north has particularly good performance when pointing to remote targets? Would pointing performance decrease as the angle of the person's misalignment from the north-facing orientation increases?

A relevant experiment to test these ideas was conducted at the Max Planck Institute of Tübingen. People who were all very familiar with the city of Tübingen were placed in an immersive virtual model of the city. While standing at a given location of the city, they were asked to point to targets that were not visible to them. Remarkably, measures of pointing error and pointing latency attested that performance was best for north-facing orientations and that accuracy decreased linearly with the angle of misalignment of the participant. Furthermore, no distance effect was found, reflecting that pointing performance was not correlated with the distance between the participant's location and the target object. The fact that a north-facing orientation benefits performance in a pointing task suggests that people may use internal representations that are based on city maps, which are typically north-oriented representations in our culture (see Frankenstein, Mohler, Bülthoff, & Meilinger, 2012).

The acquisition of cognitive maps

Cognitive maps offer a mode of representation of spatial information which is generally considered as the most evolved form of spatial cognition (see McNamara, 2013; Thorndyke & Hayes-Roth, 1982). This conception is a key feature of the developmental model of acquisition of spatial knowledge that was promoted forty years ago by Alexander Siegel and Sheldon White (1975). According to these authors, the first stage of the acquisition of spatial knowledge relative to a novel environment is limited to the *landmarks* available in this environment. In a fur ther step, people will acquire knowledge about the *routes* that connect landmarks, first in the form of non-metric representations, then progressively as representations endowed with metric properties (distances, durations, angular deviations). It is only later that a person's capacity to access a *survey* representation of the traversed environments will emerge and they will build an allocentric view of them. This conception of a "cognitive microgenetics" of space is still dominant today, even though the idea of a rigid chronology of every facet of spatial knowledge is questioned. For instance, the notion of gradual development of metric knowledge as learning proceeds tends to overcome that of development through discontinuous qualitative stages (see Ishikawa & Montello, 2006; Nadel, 2013).

Spatial knowledge builds upon a variety of sources. Exploring the environment involves sensorimotor mechanisms, of which vision and locomotion are important and must be tightly coordinated. For those people deprived of one or the other, the available modality is bound to support the bulk of spatial knowledge. One should also remember that locomotion may be implemented in either active or passive mode (when motion is ensured by a vehicle), which engenders kinesthetic conditions likely to impact representations differentially. In humans, other sources of knowledge will come in addition to (or, in some cases, substitute) actual exploration of an environment, such as map symbols, verbal descriptions, as well as computer-based positioning systems.

The dominant model of spatial knowledge acquisition recognizes that such knowledge is multimodal in nature and results from the integration of information stemming from various sources. Another feature of representations labeled cognitive maps is their gradual construction. Contrary to other forms of representation, it is unlikely for a cognitive map to be constructed instantly. Its construction rather implies repeated episodes that ensure that information gets enriched and becomes more precise, resulting in a representation containing ever more valid metric information. However, involution remains possible. A cognitive map that is not reinforced by repeated experience is likely to lose its accuracy and richness in detail. Excessively spaced or irregular visits to foreign cities give travelers the opportunity of such experiences, with the requirement to start reconstructing representations that have become too faint to be of any value.

Note

1 Further experiments in immersive virtual environments where participants explore a set of three routes four times either along the same direction or in both directions show that the participants in the unidirectional group make fewer errors and spend less time reaching the end of the paths than those in the bidirectional group. They also better recall spatial information when invited to draw the routes on a map of the environment. Access to an extra (backward) perspective may thus not always enhance route learning nor the construction of survey knowledge, especially if several routes have to be learned in the same environment. Acquiring the backward perspectives of several routes requires extra spatial processing to integrate the backward with the forward perspectives (see Zancada-Menéndez, He, Sampedro-Piquero, Lopez, & McNamara, 2016). More generally, the influence of allocentric or egocentric perspective on the process of "finding the return path" has become a crucial issue in the study of navigational performance (see Karimpur, Röser, & Hamburger, 2016).

6

MEASUREMENTS

In the two preceding chapters, we subordinated our approach to the postulate that the human mind contains entities that can be qualified as representations. As regards space, the information retrieved from these representations requires, in order to be cognitively useful, to be inscribed in a reference frame (either centered on the observer or on an alternate point selected in the environment). Furthermore, when these representations encode spatial information, they appear to be endowed with special properties that drive us to acknowledge their map-like character. It is the case, in particular, of the mental representation of configurations of objects whose survey view makes their spatial relations visible, while these relations may be hard to notice from the ground by the navigating person. We recognize that this is nothing more than a metaphor. In spite of the popularity of the concept of the cognitive map, we must ensure that only what is relevant be retained from the metaphor.

The challenge, of course, is to find adequate means to *externalize* these representations. These means are based on the collection of behavioral data according to the rules and principles of experimental methodology. These behaviors are generally taken as valid simulations of actual behaviors implemented in real environments. Sometimes, the scientist faces the necessity, for experimental convenience, to elicit and observe behaviors in artificial contexts, such as those offered by virtual environments.

Among the issues to be considered, the first one is about how people manage to recognize the place where they are standing and identify their position in the environment. A small set of objects may help for this purpose. Once the recognition process has been completed, it is possible to identify other objects in the neighborhood as well as at farther distances. People can then declare what they know or deduce from their computations as regards the direction in which the objects are, and the distance to them. People may provide such responses while they are in a stationary position. But the researcher will also be interested in the moves that people accomplish towards these objects. Proper recognition of the places traversed during the execution of moves is a prerequisite to the execution of appropriate reorientations at the right places and the right times. At a more advanced level in the construction of spatial representations based on motion, it will be necessary to investigate the mechanisms by which people update their knowledge during their progression and ensure "path integration" by combining the representations built up through successive episodes.

In all those aspects of spatial representations that are amenable to measurements, defaults or malfunctions may appear, as in any cognitive activity. These failures are sources of information of special value for the understanding of navigational processes, but also for the design of navigational aids, which must take into account the aspects by which human cognition reveals its limitations.

Places

Lynn Nadel (2013) invites us to imagine a situation where a person is transported to a particular location while blindfolded. On arrival, after removal of the blindfold, the person is asked to figure out where he/she is. In order to provide an answer, the person will first search the environment for familiar features that would make it possible to "call up" a cognitive map, the cognitive map to which those features are associated. If this is the case, the person will identify the specific surrounding environment. If such familiar features are not recognized, the person will conclude that he/she is in a completely new environment, of which no cognitive map is currently available. This experience is likely to be quite disorienting, and can even create anxiety. None of us would like to experience the feeling of being in a completely unknown environment. This situation, however, is an invitation to the person to start building mentally a map of the new environment.

What does it mean to "identify a place"? Identification is achieved when we are able to recognize a place, to identify the objects located there, but also to have a representation of its location relative to other places. The feeling of familiarity is essential in the process of identification. Golledge (1992) analyzed the processes that underlie that feeling. The first attestation of familiarity is based on a person's capacity to identify a place by naming it. A further form of attestation is provided when a person presented with a photograph of the place can testify that he/she recognizes it and declares that the photograph is veridical and that the site and the objects within it are consistent with what they are in reality. For instance, a photograph of the Eiffel Tower may trigger such recognition and naming responses. This is true for a great many people over the world, although most of them will not be able to say anything more, except that the monument is located in Paris, a piece of information available in declarative form in the knowledge base of many people, without any knowledge of its precise localization. This being said, the elementary signs of an "identification" of the Eiffel Tower do not necessarily include the *topographic* identity of the corresponding scene. This is precisely where a third dimension of place familiarity is to be introduced, that is, the capacity to report the broader geographic frame where this place is located. Identification can then be done either in an allocentric frame of reference (where is this place relative to other known places?) or an absolute one (where is it within a system of coordinates that applies to the world that contains it?).

A further measure of familiarity of a place can also be envisaged by referring to the frequency with which a person interacts with this place. Not surprisingly, the places that we are the most familiar with are those that we can best locate relative to each other.

Although place memory involves a strong visual component, verbal coding that accompanies it should not be underestimated. Route learning experiments show that depending on whether navigators are invited to remember intersections by either relying on visual impression only or by memorizing place names, the latter modality is greatly beneficial to route memory. The directions to follow from each particular place are better remembered when descriptive names are used (instead of arbitrary names unrelated to the visual impression) (see Meilinger et al., 2016).

Distances

Moving in space means to cover distances and thus implies that a navigating person accesses some knowledge of corresponding lengths. With such knowledge, the person is able to minimize the distances to traverse before reaching a target, thus to keep at its minimum the energy cost involved by moves and the associated muscular effort.

Although stated in such simple terms, the issue is made more complicated by the fact that distinct from metric distances that can be quantified objectively, we must also take into consideration the notion of *subjective distances*, also expressed quantitatively, but not necessarily correlated with actual distances. The fact remains that knowledge of distances (based on exact or approximate metrics) is the factor that offers people the possibility of making travel choices by taking into account constraints weighing on them. The efficiency of these choices is typically attested by the discovery of shortcuts or detours and, more generally, by demonstrating adaptive creativity in the resolution of navigational problems (Montello, 2009).

How are distances perceived in environments accessible to our sight, or vista spaces, as named by Montello (1993)? A classic test consists in asking people to inspect a set of objects located in an open space in front of them (at distances of between a few meters and 20 meters or so) and then asking them to travel these linear distances while being blindfolded. The distances actually covered, when people declare that they have reached the target location, are compared to the corresponding actual physical distances. Several studies have shown the remarkable accuracy of locomotor responses (Loomis, Da Silva, Fujita, & Fukusima, 1992; Thomson, 1983), provided, however, that the distances remain within certain limits and that the time elapsing between visual study of the scene and beginning of the blindfold walk does not exceed 10 seconds. Nevertheless, the fact remains that systematic distortions may affect the perception of distances. For instance, the study by Loomis and his colleagues attests that in order for sagittal (depth) and frontal (width) intervals to be subjectively perceived equal, the former must in fact be longer than the latter. However, distortions that affect distance assessment do not prevent locomotion being adequately executed.

If we now consider environments of larger size, what mechanisms allow people to acquire distance knowledge? For Montello (1997), the acquisition of such information depends on several sources, which may be redundant to some extent. The particular features of an environment provide an important informational basis, made of the events or objects met during travel. Reorientation points and intersections, and more generally all the elements distributed along the route, provide pieces of information that contribute to assessment of the route length (see Jansen-Osmann & Berendt, 2005, in an experiment using virtual reality). These sources are complemented by sensorimotor information related to the execution of travel, as well as cues provided by the perceived duration of travel and energy expenditure associated with it. Montello (2009) included in his conceptual model a mechanism consisting of summing vista distances during journeys through large-size spaces.

In addition to *direct* acquisition of information about distances, other sources can be used, such as those that make it possible to *retrieve* knowledge collected in other navigational episodes or to *infer* distances that have never been directly experienced (for instance, when we have to estimate the distance separating two cities in a country). Distance knowledge is also nourished by the experience of symbolic representations, such as geographic maps.

It remains, however, that distances encoded in human cognition are not immune to approximation, vagueness, and distortions. We review this particular aspect of distance knowledge in the next section.

Distortions

One of the classics in the study of distortions that affect memory of urban environments is a research study conducted by Richard Byrne (1979) in the town of St. Andrews, Scotland. Residents were invited to estimate by ratio scaling the walking distances between twenty pairs of well-known locations in the town. The actual distances were between 300 and 750 meters. The main findings of the study were that (a) route lengths generally tend to be overestimated, and this is still more pronounced for the shortest routes; (b) routes are judged longer if they lie within the town center; and (c) routes with several major bends are estimated longer than linear routes. These findings suggest that in the representation of environmental spaces, vector distance may not be fully preserved, even for people quite familiar with an urban environment.

Other attestations of the overestimation of urban route lengths are available (see Canter & Tagg, 1975). More particularly, routes with turns and reorientations are perceived as longer than rectilinear routes of similar lengths, and their starting and terminal points are thought to be more distant from each other (see Sadalla &

Magel, 1980; Staplin & Sadalla, 1981). This effect of route angularity suggests that travelers do not simply encode routes as concatenations of line segments (see Golledge, 1992).

More intriguing are those cases that reveal asymmetries in distance estimation. The most impressive demonstration of this phenomenon comes from the studies indicating that distances are perceived differently depending on whether they are considered between a given location and a location endowed with high cognitive saliency (as compared to a location devoid of any such feature). Sadalla, Burroughs, and Staplin (1980) proposed the idea that in an environment some places play a special role as "reference points." These are places whose locations are better known than others, more frequently visited, more visible, with important historic or cultural significance, and which are consequently more spontaneously used to locate adjacent less remarkable points. For instance, in a Parisian context, one would say that the Guimet Museum is closer to the Eiffel Tower than the Eiffel Tower is relative to the Guimet Museum. The experiment cited was conducted in a North American university campus. It confirmed that when one element of a pair of locations is more salient than the other, the distance between the reference point and the other point is systematically estimated as shorter than the distance between this other point and the reference point. This reflects that reference points in an environment contribute to the structural organization of that environment and help people to locate other points in that space. At the same time, these findings draw our attention to the fact that if the cognitive representation of distances between reference points is in accordance with a Euclidean model of space, the localization of less salient points is more consistent with the topological concept of *neighborhood*.

Perry Thorndyke (1981) conducted an enlightening study on distance estimation from cognitive maps, with a special interest for the effect of the number of intervening points along the route. Undergraduates learned the map of a fictitious environment containing twenty-one cities distributed over a network of roads (Figure 6.1). Intercity distances were multiples of 25 miles. The minimum distance between any two adjacent cities was 25 miles, and this measure was shown as a reference standard at the bottom of the map. Following map learning, one of the tasks consisted of the participants estimating for each pair of cities the distance along the shortest route connecting the pair. Responses showed that estimates increased as a linear function of the number of intervening cities along the route. The most likely explanation of this result is that people estimate distances by using a representation of the map that preserves the metric relationships among cities. They presumably scan a visual image of the route from the start point to the destination city. Scanning time along an empty road is assumed to be linearly correlated with the actual length of the route. When the scan is executed along a cluttered route, each intervening city involves a processing cost that is reflected by additional time, thus increasing the total distance estimate.

In addition to that, the representation of spatial relations among particular geographic points, such as cities, may be affected by a person's knowledge about the

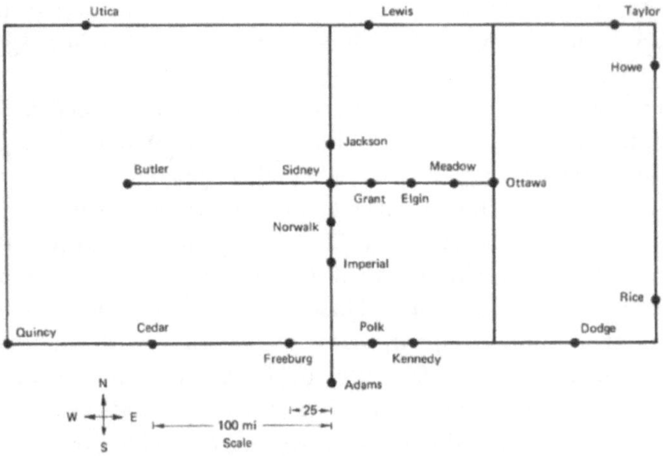

FIGURE 6.1 Map learned by the participants of the experiment. Reprinted from the article by P. W. Thorndyke (*Cognitive Psychology*, 1987), by permission of Elsevier

superordinate structures to which these points belong. For instance, the judgment about the spatial relations between two cities (which is north of the other?) is not independent from the spatial relation between the regions in which these cities are located. The most frequently cited example is that of the cities of Los Angeles, California and Reno, Nevada. The first one is thought to be west of the second, presumably due to the fact that California is overall west of Nevada, but the actual geographic relation between the two cities is different (Figure 6.2) (see Stevens & Coupe, 1978).

Furthermore, the existence of a border between the regions of which two cities are subjected to spatial comparison increases the impression of distance between the cities. When people are invited to estimate distances, they judge cities farther from each other when they belong to two distinct regions or countries than when they are located in the same region or country (see Maki, 1981; Wilton, 1979). Any sort of border thus contributes to biases in estimating distances between elements located on both sides, by enhancing the categorization process, which interferes with strict computation on the data to be taken into account. This is the case with natural boundaries (such as roads or rivers) or political borders. Emotional factors may also add to the situation and accentuate cognitive biases. For instance, fifteen years after German reunification, a task of distance estimation between two cities, one from the former FRG and the other from the former GDR, resulted in systematic overestimation of distances, as compared with estimation about two cities on the same side of the former border. This tendency was even more pronounced on the part of people who had a negative attitude toward the reunification of Germany (see Carbon & Leder, 2005). Remarkably, features of the historic past of a country can still be found in the far depths of a cognitive map.

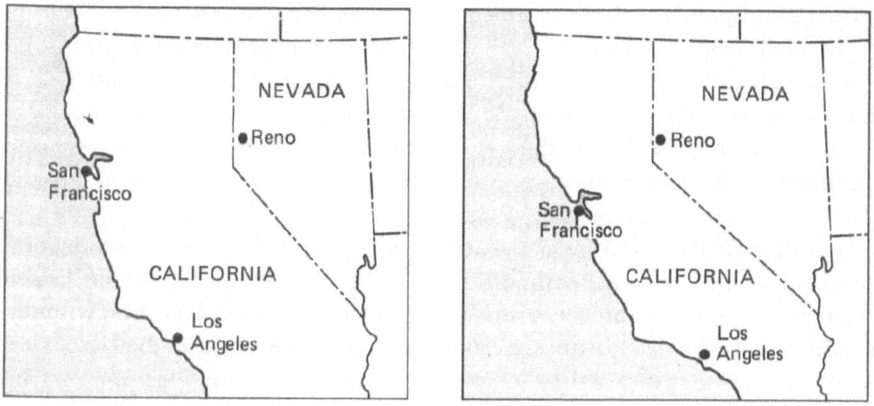

FIGURE 6.2 Left: mental representation of the localization of Reno to the east of Los Angeles. Right: actual geographic localization of Reno to the west of Los Angeles. Adapted from the book by A. L. Glass, *Cognition: A Neuroscience Approach* (2016), by permission of the author and Cambridge University Press

Angles and directions

If the localization of a site in terms of distance is so prone to error, would people be more efficient if they had to point in the direction to this site?

It is well known that angular error is greater when the place towards which we are pointing (outside our visual field) is farther to close egocentric space. From my office, I will be more accurate when pointing to the front door of the building than to the entrance gate of the campus. The amount of error is likely to increase if I have to point to the Eiffel Tower, and even still more if I have to do so towards the Statue of Liberty. Another factor to consider is our familiarity with the site that should be pointed to. Golledge (1992) underlined that pointing errors are less marked towards familiar targets than towards less well-known places.

In addition to pointing experiments in natural environments, another experimental approach consists of asking people about a real environment that they know well and inviting them to produce an equivalent of digital pointing to express directional knowledge relative to a target. A methodology has been proposed by Moar and Carleton (1982). The participants were presented with a succession of slides taken along a 240-meter route in a suburban area of San Diego, California. The route comprised three segments directly connected in succession, involving a right turn and a left turn. The slides were taken every 8 meters, thus offering a reasonable equivalent of the successive views of the route by a pedestrian. Following learning, the participants were provided with a sheet of paper. At the center of each sheet was a circle with a diameter of 5.4 centimeters. An arrow, 3.1 centimeters long, was drawn from the center of the circle and pointing up the sheet. The experimenter showed onscreen a slide taken from the route. The participants were invited to imagine that the dot in the center of the circle represented the location of the photographer when the slide was taken. The experimenter then showed a second slide, taken from another point of the route. The participants had to draw a line from the center of the circle in the direction of the location shown in the second slide (as the crow flies). This procedure made it possible to calculate the angular difference between the actual direction in the environment and the estimate given by the participant.

With this method, Moar and Carleton (1982) provided indications on the gradual decrease of directional error as route learning proceeds. They also evidenced differences between judgments made in the direction of travel and those made in a backward direction. The forward errors were significantly smaller than backward errors, indicating that spatial judgments from memory are more accurate in the direction of travel along the route than in the opposite direction. This pattern suggests that during early acquisition of a navigated environment, the spatial properties of a route are encoded in the form of sequential associations. More advanced learning is required in order for a route to be encoded as a representation endowed with allocentric properties equivalent to those of a map.

Other limitations of directional judgments can be evidenced. The most remarkable one is the non-reversibility of directional judgments for pairs of cities depending on the direction in which the judgment is to be given. Moar and Bower (1983) selected six United States cities (Chicago, Miami, New Orleans, New York, San Diego, and Seattle). All fifteen pairs of cities were considered, each of which in both directions. Thus, by reference to a given city, the participants had to estimate the direction in which each of the other five were. By using a response device similar to the one described above, it was then possible to record, for San Diego, the estimated direction to Seattle, and for Seattle, the direction to San Diego. Across the whole set of responses, the average absolute error for the two directions for a city pair was 8.0 degrees. This value reflects the non-reversibility of judgments made in both directions. It varied from 4.7 degrees (for the San Diego/Seattle pair) and 13.6 degrees (for the New York/Chicago pair). There were even people who judged Chicago as being farther north of New York than is the case, and also judged New York as being farther north of Chicago. These judgments dramatically reflect inconsistencies that sometimes affect cognitive maps.

The representation of angular sizes is highly sensitive to distortions that affect knowledge of real geographic environments. Richard Byrne (1979) asked residents of the city of Cambridge, England, to draw the junctions of well-known roads of the town. In half the cases, the roads met at 60–70 degrees, and the other half at 110–120 degrees. The analysis of drawings showed that in all cases, there were significant differences with the actual angle, and in most cases, the estimates tended to be close to 90 degrees, a confirmation of systematic bias towards the right angle in the representation of angles in urban contexts. This reminds us of the observation reported by Kevin Lynch in *The Image of the City* (1960), according to which some residents of the city of Boston describe the Boston Common as square-shaped, but also claim that walking around the park involves a total of five right angles!

FIGURE 6.3 A district of the city of Cambridge

Converging data have also been provided by Moar and Bower's (1983) study, already mentioned above. In the city of Cambridge, several districts were selected, in such a way that each of them was defined in terms of a triangle of roads (Figure 6.3). The residents used the graphic pointing device mentioned previously. For instance, they were invited to imagine that they were standing at point A (at the intersection of St. John's Street and Bridge Street), facing the direction of point B (intersection of Trumpington Street and Pembroke Street). They then had to judge the direction to point C (intersection of St. Andrew's Street and Downing Street).

Directional judgments were given for the six possible directions in each triangle. From these data, it was possible to calculate the angle for the two judged directions from a location to the other two locations in the same triad, and the same was done for the angles at the other two locations of the triad. The results confirmed that angular estimates were systematically biased toward the right angle, but the most remarkable finding was that the sum of the three derived angles for each triad was significantly greater than 180 degrees. In the example mentioned here, the real local angles were 35 degrees, 84 degrees, and 61 degrees, and they were estimated 40 degrees, 89 degrees, and 91 degrees, respectively; three figures whose total amounts to 220 degrees, a value which is obviously incompatible with the Euclidean properties of real space. The spatial information contained in the cognitive map of a familiar environment may thus appear to be rather inconsistent (however, without compromising the efficiency of people's spatial performance within that environment).

Spatial updating and path integration

Distances and directions are the basic parameters needed to determine a location relative to a reference point. Subjective estimates of these parameters may be produced from the very place where an observer is standing, but also from remote places. When they are recorded on site, the observer may be either in a stationary position or moving. The estimates that we will now consider are directly connected to a person's move (essentially locomotion in the cases illustrated here).

Our capacity to move through various environments is dependent on our capacity to identify at every instant where we are and to monitor our current position in these environments. A person deprived of the capacity of *updating* spatial information would face a real predicament. This state of affairs justifies the substantial amount of research conducted on how people update and keep track of their positions and orientations while they are traveling, as well as the positions occupied by surrounding objects (see Loomis, Klatzky, Golledge, & Philbeck, 1999; Philbeck & Sargent, 2013).

People's control over their own navigation is relatively easy when people have visual access to the environment. Perception of landmarks in the environment helps them to be aware of their position at every step of their moves. Landmarks function as beacons towards which the moving person is heading without the need to compute the parameters of the move itself. However, these computations become of primary importance when people, for some reason, have no visual perception of the environment. The concept of *path integration* (or *dead reckoning*) then appears on the scene, when translations and rotations operated by the navigating person are exploited to provide him/her with online estimation of his/her position and orientation. Internal information used to this end includes vestibular, proprioceptive, and kinesthetic signals, as well as optical flow when vision is available. However, as we will see, in most studies on path integration, the conditions are designed in such a way that participants do not have access to vision during the whole experimental session.

During the execution of a move without sight, the definition of the position and orientation of the navigating person must be specified with respect to a reference point. The most usually considered reference point is the origin point of the move. This point will have to be returned to by the navigator at the end of the locomotor episode. The representation that will be activated to this end is not a "representation of route" properly speaking. The internal representation of a route contains information pertaining to every segment covered and every turn taken along the path until the last step. In fact, the representation that underlies path integration is more abstract in nature. The task considered here does not consist in "mirroring" backwards an onward move previously executed. The final route (or return journey) relies on a representation derived from signals collected during the onward journey.

These processes have been identified in various animal species. They account for animals' homing capacities after long tortuous journeys. The most spectacular case is that of a desert ant, which, after complex navigation of about 600 meters to collect food in the Saharan desert, directly returns to the nest from which it started (Figure 6.4). A path integration strategy is invoked to account for this kind of performance (see Collett & Collett, 2000; Wehner, 2003).

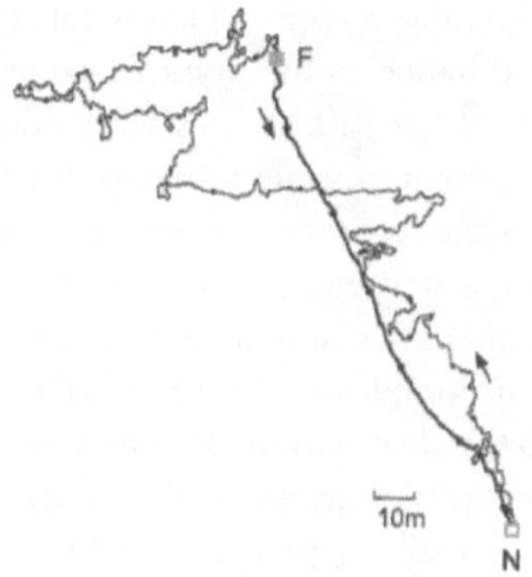

FIGURE 6.4 A desert ant's tortuous outward (foraging, from N to F) and straight homeward path (from F to N). Adapted from the article by R. Wehner and S. Wehner (*Ethology Ecology and Evolution*, 1990), by permission of R. Wehner and Taylor & Francis on behalf of the Department of Biology, University of Florence, Italy

In human experimentation, the task typically consists in placing people in a situation where they will only use information that comes from their own moves in order to update their position at each step of their progression. The experimenter will have them walk along a route while blindfolded and guided by an assistant. The route is made of a small number of rectilinear segments with a turn at the end of each segment. At the end of the journey, the navigators are asked to return to their point of departure by the shortest route (straight line), without being guided or assisted. While executing the return trip, people demonstrate that they dispose of some knowledge of the relation between the starting point and the place where they have been led (and left without any guidance). In the simplest version of this locomotor pathway completion task, the first part of the path comprises two segments, and the return path (back to the departure point) is made of only one segment. The task is thus sometimes described as a "triangle completion" task.

To take an illustrative example among the rich set of studies conducted at the University of California at Santa Barbara, we focus here on an experiment in which university students were guided while navigating along pathways of two or three segments (within a circular area with a diameter of 10 meters), and then requested to walk along the return segment to the departure point without any guidance. The results typically show that the participants' performance is – luckily – well above chance. However, performance is lower when the number of segments is greater, whether the indicator considered is the angular error of the return path (compared to what would be the correct heading) or the absolute distance error. Special difficulty appears in the case of paths in which segments intersect with each other (see Klatzky, Loomis, Golledge, Cicinelli, Doherty, & Pellegrino, 1990).

The processes involved in pathway completion remain open to various interpretations. Klatzky and her colleagues expressed their doubts about the hypothesis of continuous updating of a "homing vector" which would deliver online information about the heading and the distance to the point of origin. They favored the view that a survey representation of the path is constructed during navigation. It is this representation that would make it possible to determine the homeward trajectory, either from a survey view of the territory or from trigonometric computation.

In another study, the same task was proposed not only to blindfolded sighted participants, but also to congenitally and adventitiously blind people. The results regarding the effect of path complexity (in terms of number of segments) were confirmed. The noteworthy finding was the low performance differences between the three groups of participants, suggesting that spatial competences requested by this type of task are not really dependent on prior visual experience. Blind participants were not at a disadvantage when compared to sighted participants. As a matter of fact, it even appeared that some of the best performances came from congenitally blind participants, and some of the worst came from blindfolded sighted people (see Loomis, Klatzky, Golledge, Cicinelli, Pellegrino, & Fry, 1993). The systematic error affecting final performance does not seem to be due to errors taking place during progression along the last segment, but to errors that affect the memory encoding of information collected along the segments covered with guidance (see Fujita, Klatzky, Loomis, & Golledge, 1993; Klatzky, Loomis, & Golledge, 1997).

Updating one's position during a journey seems to be governed by automatic processes to a large extent. The coding mechanisms brought into play while progressing along the initial part of the journey are only moderately sensitive to distracting sources such as concurrent irrelevant verbalization (repeating a sequence of nonsense syllables) or some activity involving higher cognitive load (counting backwards from a given number). On the contrary, executing irrelevant movements that interfere with locomotion (for instance, being instructed to move rightward or backward a few meters) results in significant performance decrease. During locomotion, the person cannot ignore these interfering movements, which are disruptive to the path integration process (see May & Klatzky, 2000).

The uncomplicated nature of the paths that form the main material of experiments on path integration forces us to question their ecological representativeness. There are good reasons to think that spatial updating processes investigated for several decades should be extended to more complex situations. In particular, it is well established that path integration becomes much more difficult when a traveler moves along curved segments (see Cornell & Greidanus, 2006). On the other hand, there exist experimental situations where path complexity has no effect on performance (see Wiener & Mallot, 2006). These results force researchers to question the postulate that complex natural paths would be equated to plain combination of short rectilinear segments (as those which constitute the basic material of research on path integration).

PART III

Brain and sensorimotor systems

Functions and dysfunctions

Psychology is driven by the requirements of the experimental method. Knowledge of the mechanisms that generate behavior more and more relies on the development of neuroimaging techniques. Thanks to neuroimaging, the relationships between cognitive neuroscience and psychology have profoundly evolved. In his book on "the making of ideas," the late Marc Jeannerod (2011) underscored that as soon as neuroscience became cognitive, it got closer to psychology. This occurred in such a way that nowadays the two disciplines can no longer progress in isolation from each other. It is common today to see the same scientists committed to accounting for both the mental and cerebral levels of cognitive processes in a single integrated action. More than one example will be provided in the following chapters, illustrating concerted elaboration of models of mental processes and their neuronal infrastructure.

We will first review the brain structures that have been identified for their role in the processing of spatial information. Neuropsychological observations, now complemented by data provided by neuroimaging, have incrementally drawn the picture of the "spatial brain." A great deal of effort has been devoted to account for the role of the hippocampus as the brain region where cognitive maps are elaborated. But the hippocampus is only one component of a broader network responsible for the representation of spatial information and the planning of navigation.

The brain is a vulnerable structure, subject to potential damage, some types of which electively affect spatial cognition and navigation. Impairments due to focal brain injuries or (normal or pathological) aging frequently result in disorientation, a severe deficit that impacts people's environmental and social integration.

Besides, a number of sensory, cognitive, or motor impairments affect individuals' capacity to entertain satisfying relationships with their spatial environment.

Of particular significance are the situations where sensory systems that support space representation are partly or totally unavailable. Vision occupies a very special place in this context. Not only the investigation of blind people's brain activity is likely to inform us of the mechanisms involved in the representation of space, but from a practical perspective abundant research is also being devoted to the design of navigational aid systems for the blind.

7

THE SPATIAL BRAIN

Research on brain representation of spatial knowledge has always entertained close connections with cognitive theory. In particular, special interest has been expressed in how spatial frames of reference are represented in the human brain. Neil Burgess (2006) suggested that the distinction between egocentric and allocentric reference frames superposes itself to the distinction between the role of dorsal and ventral paths in visual processing, as established originally by Ungerleider and Mishkin (1982). Interactions between the two systems are attested by the role of the parietal cortex in creating egocentric representations, which then serve as inputs for creating allocentric maps in the hippocampus. Another form of cooperation, symmetrical to the preceding one, is illustrated when allocentric representations in a person's long-term memory are activated to create egocentric representations designed to serve a specific navigational task.

Whatever the case, since the early 1970s the most critically attested brain structure responsible for the representation of spatial information is the hippocampus, with which we start our review of the components of the spatial brain.

The hippocampus

The hippocampus is a structure that receives messages from many regions of the brain, and in turn sends messages towards several cortical and subcortical regions (Figure 7.1). It is implied in other non-spatial functions. It has been suggested to play a role in information integration, as well as episodic memory. The fact remains, however, that the hippocampus is the core structure responsible for many aspects of spatial cognition, while not being the sole brain region concerned, as will be made clear in the rest of this chapter.

O'Keefe and Nadel proposed considering the hippocampus as the locus of encoding of allocentric spatial information in the form of "cognitive maps."

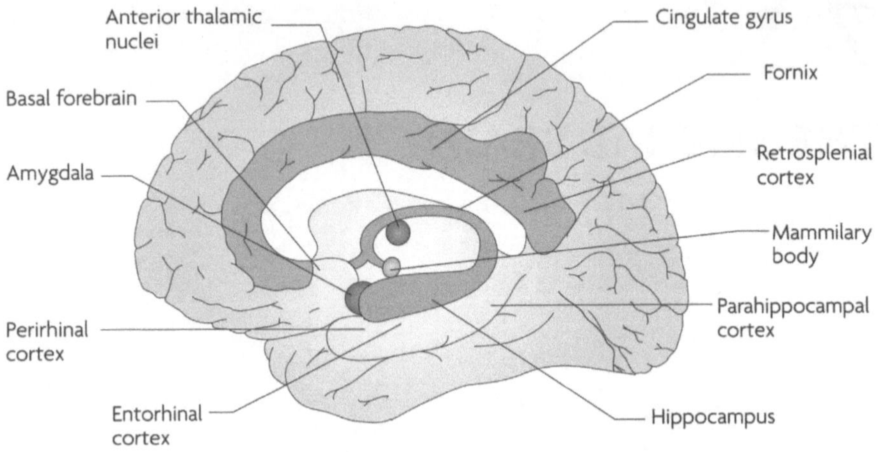

FIGURE 7.1 The hippocampus and the brain structures to which it is connected. Reprinted from the article by C. M. Bird and N. Burgess (*Nature Reviews Neuroscience*, 2008), by permission of N. Burgess and the Nature Publishing Group

This is the meaning conveyed by the title of their book, *The Hippocampus as a Cognitive Map* (1978).

The starting point of the story dates back to the recording of single-unit neuronal activity in the hippocampus of the freely moving rat and the discovery by O'Keefe and Dostrovsky (1971) that some neurons are specifically activated when the animal is at particular places in its environment. The so-called "place cells" are found to fire maximally when the animals occupy specific regions of the testing platform. Further recordings confirmed that place cells are associated to particular "place fields." When the animal leaves these places, the corresponding neurons become inactive.

Remarkably, on multiple exposures of the animal to the same environment, the place cells reinstate the same activity, but when the animal is immersed in a novel environment, the same cells are activated for new place fields, which results in the creation of a new firing map. Such remapping suggests that the networks of place cells are not fixed, but on the contrary are dynamic, with new connections generating new configurations.

The firing rate of place cells typically increases as the rat traverses specific regions of its environment. This finding provides a striking example of the connection between a neuronal activity and a basic spatial behavior. It is on these grounds that O'Keefe and Nadel (1978) developed the concept of the hippocampus as responsible for the formation of cognitive maps.

It took more than twenty years for researchers to collect evidence of the fact that the human brain exhibits similar features to the rodent's brain. When people are invited to explore a city through a virtual environment, recordings of single neurons in the hippocampal and parahippocampal regions similarly reveal that neurons respond when participants are at specific places in the environment. Place-responsive cells appear to be more prevalent in the hippocampus than in any other part of the brain (see Ekstrom et al., 2003).

Further research programs extended our knowledge of those neurons that exhibit particular response patterns in relation to an animal's position in a spatial environment (see Poucet et al., 2004). Other types of neurons were then identified, in particular those whose corresponding fields show a regular spatial distribution by forming equilateral triangles. The so-designated "grid cells" were found in the entorhinal cortex, a structure of the medial temporal lobe which projects to the hippocampus. Grid cells are thought to encode information that is relevant for path integration (see Hafting, Fyhn, Molden, Moser, & Moser, 2005; Moser, Kropff, & Moser, 2008). This complex of cortical microcircuits participates in a broader network dedicated to the representation of self-location. More recently, experiments involving human participants succeeded in collecting neuroimaging signals that were consistent with the presence of grid cell activity during goal-directed navigation and planning, as well as imagined navigation (see Horner, Bisby, Zotow, Bush, & Burgess, 2016).

The scientific advances that documented the neuronal contribution to the representation of space received special acknowledgment when John O'Keefe, May-Britt Moser, and Edvard Moser were awarded the Nobel Prize in Physiology or Medicine in 2014 "for their discoveries of cells that constitute a positioning system in the brain" and thus form a kind of internal map of an environment.

In humans, as also demonstrated in various animal species, a strong attestation of the role of the hippocampus in the creation of cognitive maps has come from the accumulated neuropsychological evidence that hippocampal lesions impair human spatial memory by affecting the strategy that relies on the exploitation of visual landmarks (see Bohbot, Iaria, & Petrides, 2004; Spiers, Maguire, & Burgess, 2001; Wolbers & Hegarty, 2010).

In humans specifically, one of the most impressive demonstrations of the role of the hippocampus in space representation arises from the investigation of people identified as "experts" in spatial navigation. Taxi drivers, who have to make navigational decisions repeatedly in complex environments, belong to that category.[1] In some cities, in particular London, they have the additional characteristic of being submitted to extensive training for several years before becoming licensed drivers and being allowed to exercise their profession. More than other people, London taxi drivers may be considered as experts in spatial representation and navigation.

Neuroscientific investigations by Eleanor Maguire at the Institute of Neurology in London are particularly informative in this respect. By using positron emission tomography (PET), Maguire and her colleagues evidenced activation of the right hippocampus in drivers submitted to tasks calling for their topographic memory, such as imagining and describing routes through the city of London (see Maguire, Frackowiak, & Frith, 1997). Later on, their investigation consisted of identifying whether the brain of these "professionals of space" exhibited any special characteristics. Structural magnetic resonance imaging (sMRI) showed that the posterior part of

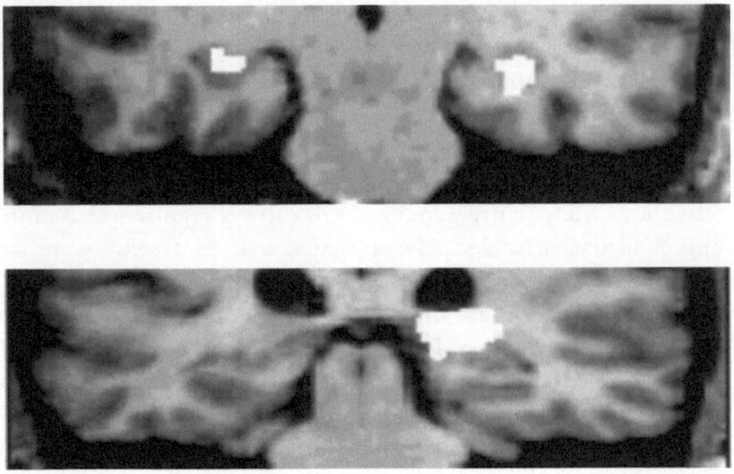

FIGURE 7.2 Above: increased gray matter volume in the posterior hippocampi in taxi drivers compared with control participants. Below: area in the right posterior hippocampi where the longer the time spent taxi driving, the greater the gray matter volume. Reprinted from the article by E. A. Maguire, H. J. Spiers, C. D. Good, T. Hartley, R. S. J. Frackowiak, and N. Burgess (*Hippocampus*, 2003), by permission of E. A. Maguire and John Wiley and Sons

their hippocampus was bilaterally larger than that of a control population, a finding that supported the hypothesis that the volume of gray matter in this region is an acquired characteristic that results from thorough spatial learning to which these people were submitted (Maguire et al., 2000). In a still more convincing way, the increased size of hippocampus was not found in drivers of other vehicles, such as bus drivers, who drive repeatedly along preset routes and therefore do not have to develop any form of spatial planning. Furthermore, in taxi drivers, the hippocampus is all the more developed as these professionals have longer experience in driving (Figure 7.2) (see Maguire, Spiers, Good, Hartley, Frackowiak, & Burgess, 2003). These findings provide an interesting example of cerebral plasticity dependent on learning and exercise.

It is relevant to note that the same group of researchers in London demonstrated that the specific activation of the taxi drivers' hippocampus not only takes place when they are participating in simplified experimental tasks in the lab, but also when they are immersed in a realistic reconstruction of conditions of their professional activity, where the navigation task develops in parallel, and even in competition, with many other perceptual and cognitive solicitations (see Spiers & Maguire, 2006).

One further step in the investigation of the role of the hippocampus was taken when another research group strived for demonstrating that the activation of this structure is related to the use of particular strategies. The demonstration is based on a situation in which participants first learn to execute moves within a virtual environment where several visual landmarks are located. Starting from a behavioral analysis and an inventory of post-experimental comments, it is possible to make a distinction between two groups of people, namely, those who adopt a strategy that consists of using visual landmarks in order to repeat a memorized route, and those who prefer a strategy based on the memorization of a sequence of moves (take the first segment to the left, then the next one to the right, etc.). A functional magnetic resonance imaging (fMRI) study revealed hippocampal activation in those people who use landmarks as navigational guides, contrary to those who rely on the sequence of turns, for whom the caudate nucleus was activated (Iaria, Petrides, Dagher, Pike, & Bohbot, 2003).

Morphometric analysis of the two brain regions is highly informative. It attests that the density of gray matter is higher in the hippocampus of people who prefer to use spatial landmarks (and whose performance is lower when these landmarks are removed from their sight), whereas density is lower in the caudate nucleus. The probability of a person exploiting the spatial landmarks is higher when the gray matter density is higher in the hippocampus, while the probability of using memory of reorientations is higher in people whose gray matter density is higher in the caudate nucleus (Bohbot, Lerch, Thorndycraft, Iaria, & Zijdenbos, 2007). These findings suggest a "competitive interaction" between the two brain regions, which apparently support distinct strategies, but jointly remain at the service of the same objective, that is, effective navigation.

While the hippocampus is recognized as the main site of cerebral representations of space and has been proposed as the best candidate to store the representation of our spatial environments, it is also established that it is involved in episodic memory (see Dudchenko, 2010; Redish & Ekstrom, 2013). It remains in any case the structure most closely associated to the notion of allocentric representation.

The parahippocampal cortex

Other regions in the vicinity of the hippocampal formation have also been identified for their role in spatial representation and behavior. In particular, PET and fMRI have revealed that the parahippocampal gyrus is activated during the learning of topographic information, but also when this information is reactivated in memory, whether the learned environment was a real one (Ghaëm et al., 1997) or a virtual maze (Aguirre, Detre, Alsop, & D'Esposito, 1996). The posterior part of the parahippocampal region responds strongly to visual scenes such as landscapes and cityscapes (Epstein & Kanwisher, 1998), and lesions of this region are known to engender topographical disorientation (Epstein, 2008).

In this respect, the works by Gabriele Janzen, in Nijmegen, brought original information by revealing that the parahippocampal gyrus responds differentially to the visual landmarks that are particularly relevant for wayfinding and those that do not have such function. The experiments consisted in having people learn a navigational episode in a virtual environment where objects (toys) were located close to reorientation points. Two cases were considered (Figure 7.3). Some of these

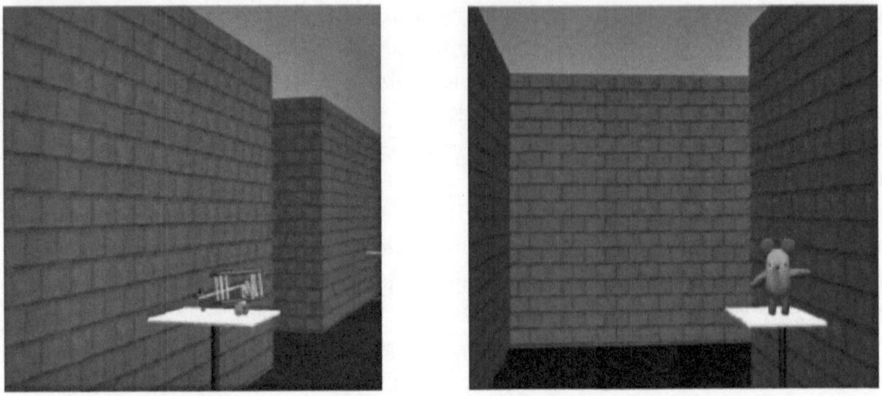

FIGURE 7.3 Left: object at decision point. Right: object at non-decision point. Reprinted from the article by G. Janzen and M. van Turennout (*Nature Neuroscience*, 2004), by permission of G. Janzen and the Nature Publishing Group

objects were placed at points where a reorientation decision had to be made (here, for example, the airplane was at a place where the navigating person had to choose between two directions, either to proceed straight on or turn left). Other objects were navigationally irrelevant (here, the puppet was at a point where a right turn is mandatory and does not call for a decision).

After learning, the participants were presented with a recognition task involving previously seen objects and distractors. The analysis of fMRI recordings revealed an increase in activity of the parahippocampal gyrus for objects which had been seen at decision points, as compared to objects devoid of any function of aid for reorientation. Furthermore, the cognitive differentiation of objects as a function of their navigational relevance was rapidly induced (as early as the participants' first exposure to the environment), and appeared to be a long-lasting phenomenon (still observed after a 24-hour interval). It is also noteworthy that at rest, the functional connectivity between the parahippocampal gyrus and the hippocampus was correlated with people's self-rated navigational ability (see Janzen, Wagensveld, & van Turennout, 2007; Wegman & Janzen, 2011). Remarkably, a study combining voxel-based morphometry and functional connectivity also attested that the gray matter volume of the hippocampus, parahippocampus, retrosplenial complex, and entorhinal cortex is correlated with people's self-reported navigational ability and sense of direction (Hao et al., 2016).

The parietal cortex

We now look at another brain region involved in the processing of space. The parietal cortex, more particularly its posterior part, is considered to be critical to people's knowledge of their position and the positions of other objects in an environment. It has long been established that the left posterior parietal cortex is important for the computation of egocentric perspective transformations (see Zacks, Rypma, Gabrieli, Tversky, & Glover, 1999). If the parietal cortex is a likely support to the egocentric processing of space, as we will see, it could also be the locus where functional interactions between egocentric and allocentric frames of reference take place, and where representations issued from the two systems are integrated (see Kesner & Creem-Regehr, 2013).

In Chapter 5, we reported findings from a study on judgments of relative distances (Committeri et al., 2004). The study revealed behavioral similarities, and by inference similar processing cost, in a task where judgments were made in either egocentric or allocentric conditions. In the first case, judgments were made about distances between oneself as an observer of a scène and an object in the scene. In the second case, judgments were made about distances between two objects or between an object and a building. Here, fMRI was used to measure brain activity associated with the various types of spatial coding. The activation of a parieto-frontal circuit appeared to be the main characteristic of egocentric (viewer-centered) coding, involving posterior parietal cortex and frontal premotor regions bilaterally, but more extensively on the right. The allocentric conditions, and more particularly the condition centered on the built environment, activated the same parieto-frontal network as the egocentric condition, but also the ventral regions bilaterally. This activation of the ventromedial occipito-temporal cortex included the fusiform and lingual gyri, as well as to be related to the point noted above – the posterior parahippocampal region. Further activation was observed in the retrosplenial cortex (a structure that sends inputs to the parahippocampal gyrus and, through it, to the hippocampus). The significant information was that beyond a few differences between the two allocentric conditions, a strong contrast was established between these conditions and the egocentric one. The allocentric perspective clearly involved ventral regions in addition to the parieto-frontal network, whereas only the latter was concerned by the egocentric perspective.

Further works by Tina Iachini and her colleagues in Naples brought valuable confirmatory information. Their study was about patients who had a parietal lesion (but no sign of unilateral spatial neglect). The experimental situations were similar to those of Committeri and her colleagues (2004), except that the judgments here were given from memory, a few seconds after the presentation of the scene. The study showed that patients with right brain damage exhibited special difficulty in the egocentric condition, while those with left damage experienced more difficulty in the allocentric condition. To summarize, the results suggested that the right hemisphere was specialized in the processing of visuo-spatial information according to an egocentric frame of reference, while a left hemisphere lateralization was confirmed, although less markedly, for allocentric processing (see Iachini, Ruggiero, Conson, & Trojano, 2009).

Interactions

Once the distinction between egocentric and allocentric frames of reference is acknowledged, the question arises of their association with specialized brain structures.

Some of the data reported above provide hints to such associations. This approach, however, does not preclude that common networks of brain areas may be recruited by the two types of spatial processing. For instance, after learning of an environment by locomotor exploration along a route or by study of a spatial map, PET recordings reveal that memory reconstruction of the environment (in the form of mental images) is associated with right hippocampal activation in both cases. Activation of the parahippocampal gyrus, on its part, is only found after route learning (see Mellet et al., 2000). When participants are scanned with fMRI *during* the learning phase (while attending a navigation movie taken from the ground-level or from an aerial perspective), it also appears that both perspectives recruit a common network, although some differences are noted. While route encoding recruits temporal and parietal regions that are not activated by survey encoding, the latter recruits a subset of areas recruited by the former. Note that if route and survey encodings lead to different activation patterns, such differences should not necessarily be taken as reflecting separate learning systems, but rather differential activation within the same brain system. Route and survey encodings are thus thought to recruit a common system in the brain, with survey encoding recruiting a specialized subset of route-encoding areas, and without any evidence of survey-specific activation (Shelton & Gabrieli, 2002).

In Chapter 5, we considered the respective features of egocentric and allocentric representations, and we referred to the situations in which a person would have to switch from one frame of reference to another. This process is likely to occur when people update their position while traversing an environment.

Using fMRI, evidence is available that the hippocampus and the retrosplenial cortex are involved in topographical orientation. When people are moving within familiar surroundings, spatial updating is required when they adopt another frame, whether this switching consists of translating egocentric information in the form of a survey cognitive map, or turning from a map to the frontal view of the path followed (see Iaria, Chen, Guariglia, Ptito, & Petrides, 2007; Wolbers & Büchel, 2005). The special role of the retrosplenial cortex is supported by the works that have explored the concept of dual strategies combining egocentric and allocentric perspectives (see Latini Corazzini et al., 2010). It is also suggested to be involved in a critical aspect of navigation, that is, when people have to establish their position and orientation relative to fixed elements of the external world (see Marchette, Vass, Ryan, & Epstein, 2014).

People's capacity to adopt either egocentric or allocentric perspectives and to switch from one strategy to the other from trial to trial has been mentioned in Chapter 5, whether this happens spontaneously or on request (see Iglói et al., 2009). The same behavioral paradigm was implemented by the same authors and extended by collecting fMRI measures. Recordings attested for the connection between use of egocentric representations and activation of the left hippocampus, whereas the right hippocampus is activated when allocentric representations are brought into play. In sum, both hippocampi would support representations that complement each other and are both helpful to navigation, namely, representations that code temporal sequences (on the left) and those concerned by places and their configurations (on the right) (Iglói, Doeller, Berthoz, Rondi-Reig, & Burgess, 2010). These data are consistent with the hypothesized role of the hippocampus as reported above.

Cognitive maps and the brain

In this chapter, we have reviewed the main brain regions identified for their contribution to spatial representation and memory. Among these regions, the hippocampus definitely appears to be a critical one, as the cerebral infrastructure in charge of encoding cognitive maps (O'Keefe & Nadel, 1978). The concept of the hippocampus as a site for mechanisms that ensure allocentric representation of space is supported by a further study which interestingly used a measure of the "adaptation" of fMRI signals. The phenomenon of adaptation corresponds to the reduction of cerebral response when an item or elements of an item are repeated. This reduction is interpreted as reflecting the overlap between representations of the first and second item. In the experiment reported here, volunteers were presented with photographs of ten prominent visual landmarks (buildings, statues, etc.) from a university campus. Twenty-two distinct photographs were taken of each landmark (from various points of view), resulting in a total of 220 photographs, which were presented in succession.

As expected, when several different views of the same landmark are shown successively, an adaptation effect is evidenced (in the hippocampus, the parahippocampal gyrus, and the retrosplenial complex). But what is also demonstrated is an adaptation effect between pairs of distinct landmarks (in the left anterior hippocampus), which varies as a function of the real-world distance between successively viewed landmarks. Thus, the cerebral structure that is supposed to encode map-like representations exhibits greater adaptation (lower fMRI response levels) when close landmarks are shown in immediate succession, and less adaptation (higher fMRI response levels) when the presentation of a given landmark is followed by a distant other one (see Morgan, MacEvoy, Aguirre, & Epstein, 2011).

To summarize, when people, familiar with an environment, view photographs of landmarks of that environment, the hippocampal response to each landmark depends on the actual distance that separates a landmark from the one seen just before. This finding indicates that the hippocampus considers locations that are physically closer in space to be more representationally similar, and more distant locations to be more representationally distinct. The hippocampus thus exhibits a key feature of a cognitive map, that is, to preserve real-world distance relationships. Importantly, the distance effect is observed as participants are not engaged in a navigation task, but are merely invited to view and covertly identify landmarks. These data are in good agreement with the idea that the hippocampus is the cerebral support of the cognitive map of an environment.

Note

1 Taxi drivers have received special consideration on the part of a good number of researchers in psychology (see Chase, 1983; Kalakoski & Saariluoma, 2001; Pailhous, 1970; Péruch, Giraudo, & Gärling, 1989).

8
WEAKNESSES

As the whole cognitive sphere, spatial knowledge is subject to a variety of dysfunctions. In Chapter 6, we considered examples that can be qualified as commonplace, reflecting approximations and biases that affect people's daily spatial computations. But our spatial knowledge and sense of direction are also threatened by accidents that may affect the cerebral infrastructure, an extremely fragile object, if any. The ensuing dysfunctions lead to weakened abilities in spatial orientation, even in very familiar environments. From topographical disorientation caused by neurological damage to anxiety or phobic disorders, the pathological pictures are quite diverse. The coming pages can only give a summary of them.

Topographical disorientation

Topographical disorientation is a deficit that affects a subset of space-related behaviors. It is reflected in a person's inability to find his/her way in an environment and progress in it in an adaptive manner. The most severe cases are those where spatial disorientation is associated with focal brain damage. Their study helps researchers to account for the specialization of particular brain regions in representing an environment and moving through it. The terminology ("topographical disorientation") is somewhat vague if one considers that the detected behavioral disorder may result from damage impacting any of the particular sub-capacities whose combination enables a human to navigate. Losing control of direction may occur in a person whose capacities of perception and representation of an environment are intact, while other forms of disorientation are due to reduced (or lost) capacity of identifying places or landmarks (a form of agnosia).

As a consequence, topographical disorientation cannot be described as a homogeneous disease. Facing the diversity of its forms and origins, researchers have developed efforts to propose taxonomies of topographical disorientation (see Aguirre & D'Esposito, 1999). As a starting point, a differentiation is made between a patient's inability to locate objects in space from an egocentric perspective, and his/her incapacity to understand directional information conveyed by objects (associated to parietal and cingular lesions, respectively). Inability to identify landmarks is typical of lesions affecting the lingual gyrus. These forms of disorientation are observed in environments that a person is familiar with, but a more specific form may be revealed in novel environments. Deficits that affect the acquisition of new topographical knowledge are typically correlated with parahippocampal lesions.

The functional significance of the hippocampus in the spatial domain, and more particularly in the building of allocentric representations during spatial learning, is attested by the impact of lesions to this structure in a large range of disorientation phenomena. Researchers, nevertheless, remain confronted to the fact that although spatial orientation deficits are correlated with clear lesion sites, at least a part of spatial information is still accessible to patients without a hippocampus. This is attested by the numerous analyses of the HM case, a patient examined by generations of neurologists and neuropsychologists until his later years (see Corkin, 2002; Dudchenko, 2010).[1]

There are several indicators of the difficulties experienced by hippocampally compromised patients when they have to navigate in a new environment. A remarkable case is that of a London taxi driver with a bilateral hippocampal lesion, who had intact memory of the streets and buildings of the city of London through which he had moved for forty years, but was unable to move by himself in a new urban environment (see Maguire, Nannery, & Spiers, 2006). Another impressive case is that of a patient who had suffered an acute ischemic right posterior hippocampal lesion and demonstrated relative preservation of his cognitive functions involving orientation and language, while displaying impaired abilities of autonomous navigation (see Aradillas, Libon, & Schwartzman, 2011).

The impact of right hippocampal lesions on the learning of new routes is amply documented (see Barrash, Damasio, Adolphs, & Tranel, 2000). It is also attested in spatial tasks that require path integration. Patients who underwent unilateral temporal lobectomy with resection of the amygdala and two thirds of the hippocampus were examined in the paradigm of return to the starting point after walking along two segments while blindfolded (the triangle completion task described in Chapter 6). Patients with right lobectomy were impaired at estimating the turn required to walk along the third segment and return to the starting point, which was not the case for patients with left lobectomy. Note that their impairment did not affect the length of their trajectory along the third segment, but its angular direction. The research confirmed that the lateralization of the spatial function under study, well established from the classic situation where disoriented patients are tested in real visual environments, is also observed in the non-visual situation where patients can only encode information from self-motion (idiothetic) signals (see Worsley, Recce, Spiers, Marley, Polkey, & Morris, 2001). The same lateralization effect has been confirmed in a number of similar locomotor tasks (see Philbeck, Behrmann, Levy, Potolicchio, & Caputy, 2004).

To make the picture more complex, we must note that fMRI detects the activity of other brain regions in navigation tasks, such as the prefrontal cortex, particularly its ventromedial and ventrolateral parts (see Spiers & Maguire, 2006). This localization is confirmed by the case of a patient with such a bilateral ventromedial lesion, unable to travel in his own city, while keeping correct memory of the details of his environment. The tasks to which the patient was submitted made it clear that his deficit mainly resulted from his inability to maintain in memory his goal destination, with the consequence that while walking along a route, he was inevitably "attracted" to more familiar places. The diversion of attention from the goal while executing a task, even a task that is not spatial in nature, is typical of patients suffering such prefrontal lesions (see Ciaramelli, 2008).

One must also mention those cases of spatial disorientation not associated with any identifiable neurological alteration. In the absence of brain damage or malformation, some people report their incapacity to have a mental representation of their environment and the serious deficits in orientation that they experience in unknown spaces, but also familiar environments (their home, their neighborhood). They demonstrate such difficulties from their childhood, although the disorder is not accompanied with other cognitive or sensory troubles, nor laterality confusion. The fMRI analysis of a particular case does not reveal any activity of the hippocampus or the retrosplenial cortex when the patient is invited to mentally represent a familiar space. The possibility remains that this specific deficit encompasses a heterogeneous set of functional disorders, as is the case for disorientation associated with brain damage (see Iaria & Barton, 2010; Iaria, Bogod, Fox, & Barton, 2009).

Disorientation and cognitive aging

Among the cognitive functions whose decline with age is well documented, spatial capacities occupy a special place (see Techentin, Voyer, & Voyer, 2014). As regards spatial orientation and memory for new environments, an illustrative study is to be mentioned here. It involved a sample of eighty participants between the ages of 18 and 78 years, with varied educational backgrounds, without any history suggestive of a neurological condition. The study revealed that in learning of a 500-meter route through a medical center, difficulty (i.e., total number of errors across repeated trials) increased monotonically as a function of age. When measure of performance concentrated on the number of errors at intersections, decline was evidenced in the group aged 60 to 69, and it was particularly marked beyond 70 (see Barrash, 1994).

The interpretation of these results needs some caution if one is inclined to attribute the observed phenomenon to the decrease of a capacity concerning electively the processing of spatial information. We must consider the decline that more generally affects memory functions, attentional capacities, and executive control (as well as perception of visual flow). While elderly people produce more errors than younger people when reaching an intersection, they are as good as their counterparts at recognizing visual landmarks along a route. But they are not as good at memorizing the temporal order in which they have passed these landmarks (see Wilkniss, Jones, Korol, Gold, & Manning, 1997). Incidentally, the vulnerability due to aging as regards learning of new environments and moving through them is naturally expressed by the fact that the elderly are inclined to avoid unfamiliar routes in natural circumstances of autonomous navigation.

The recourse to simulated navigation situations in digital environments has been spreading in the last few years, in particular in studies involving aged people. Not only do virtual reality techniques allow experimenters to control the content of the environments, but they also ensure for vulnerable participants a level of safety that is never fully guaranteed in real environments. These factors explain why virtual reality has become so popular for experimental research.

Such approaches, which take over the experiments conducted in natural environments, contribute to evidence the effects of cognitive aging on (virtual) navigation and the construction of cognitive maps (see Jansen, Schmelter, & Heil, 2010; Liu, Levy, Barton, & Iaria, 2011; Plancher, Gyselinck, Nicolas, & Piolino, 2010; Taillade et al., 2013). However, these techniques have their limits, the most troublesome of which is that they do not provide participants with proprioceptive feedback which is an integral part of locomotor experience in normal circumstances. In particular, the absence of vestibular information may render the control of spatial behavior more difficult for the elderly. Their difficulty in encoding information related to movements is explained by the limits imposed on them in the processing of vestibular and proprioceptive information (see Moffat, 2009).

In contrast, the contribution of kinesthetic and vestibular information is essential in the classic path integration tasks, as reported in Chapter 6. In the paradigmatic version of the triangle completion task, the elderly's performance along the third segment is not quite different from that of younger people, but the situation changes when the participants are passively transported along the first two segments in a wheelchair. In this case, they miss every kinesthetic signal and can only rely on vestibular information. When moving along the third segment, performance of the older participants (in their seventies) deteriorates more markedly than that of younger ones, both as concerns distance and angle. It remains, however, that sensitivity to vestibular information regarding acceleration and rotation declines significantly in late adulthood (see Allen, Kirasic, Rashotte, & Haun, 2004).

Deficits affecting the acquisition and use of new spatial knowledge by aging people must be related to the decreased activity of the concerned brain regions: the hippocampus, the parahippocampal gyrus, and the retrosplenial cortex. Data available from fMRI recordings attest that during spatial learning, the hippocampus of older adults (in their seventies) exhibits a lower level of activation (although there is no hippocampal volume differences with the younger). On the other hand, the parahippocampal and prefrontal functional attenuation is accompanied by significant volumetric reduction (see Antonova, Parslow, Brammer, Dawson, Jackson, & Morris, 2009). It might also be the case that hippocampal functions are affected by aging. In younger people, the hippocampus does play the role that is generally assigned to it in coding geometric information about an environment (by contributing to the elaboration of cognitive maps). For older people (by 60 years), the hippocampus would be more particularly in charge of the coding of local cues (such as visual landmarks) (see Schuck, Doeller, Polk, Lindenberger, & Li, 2015).

Neurodegenerative diseases

Normal cognitive aging may sometimes turn into more severe neurodegenerative pathologies. Mild cognitive impairment (MCI), which is generally considered as prodromal expression of pathological aging, takes the form of an aggravation of the functional impairments typical of normal aging, those concerning primarily the hippocampus and the prefrontal cortex. Spatial disorientation is one of the markers that flag people with the highest risk of developing dementia. In Alzheimer's disease (AD), all the brain structures responsible for spatial representation and navigation suffer from more or less severe alteration. Allocentric representations are the first to be affected and as the disease worsens, route representations are damaged as well (see Lithfous, Dufour, & Després, 2013). The difficulties affecting route planning have been amply documented (see Passini, Rainville, Marchand, & Joanette, 1995).

For AD patients, topographical disorientation is first due to their difficulty to merely recognize the places where they are standing (although cases of disorientation without visual agnosia may also be found). Neil Burgess and his colleagues in London have reported the case of a patient still at the beginning of her disease and showing severe impairment of her displacements. Her visual memory of landmarks and scenes was preserved, but her deficit was restricted to an incapacity to locate the position of objects in the environment from memory. When keeping the same (egocentric) point of view on a scene, the patient was able to remember the position of objects within it. But when she had to take another point of view on the scene, she was no longer able to retrieve the relative position of the objects. This impairment of allocentric memory of an environment may account for the difficulties experienced by the patient in navigational tasks (see Burgess, Trinkler, King, Kennedy, & Cipolotti, 2006).

Neurological investigations shed light on the brain damages that cause orientation and navigation impairments. In a particularly illustrative study, both MCI and AD patients were tested in a route-learning task. While all were efficient at recognizing landmarks, about half of them got lost on the route. Regardless of their diagnosis group, the patients who got lost had lower right posterior hippocampal and parietal volumes than those who succeeded on the navigation task (see delpolyi, Rankin, Mucke, Miller, & Gorno-Tempini, 2007). Navigation tests are considered nowadays as bringing early relevant information in the diagnosis of AD (see Bellassen, Iglói, Cruz de Souza, Dubois, & Rondi-Reig, 2012).

Lastly, the measurement of spatial orientation capacities of neurological patients, as is also the case in the study of normal cognitive aging, recourses more and more to virtual environments, with the argument of satisfactory validity in comparison with experiments in real environments (see Cushman, Stein, & Duffy, 2008; Morganti, Stefanini, & Riva, 2013).

Spatial neglect

Unilateral spatial neglect occupies a very peculiar place among space-related pathologies. The syndrome is characterized by a deficit of patients' attention or interest for information present in one of their visual hemifields, that is, to report about it verbally, to react by motor response, or to orient themselves towards it. Patients behave as if one half of their pericorporeal space no longer existed. Such neglect (more appropriately termed "hemineglect" or "unilateral neglect") frequently results from a lesion of the contralateral hemisphere, typically in the parietal lobe. The most frequently described form is left hemineglect, which occurs in patients whose right parietal cortex has been damaged (see Bartolomeo & Chokron, 2002; Danckert & Ferber, 2006).

The neglect syndrome is shown by people whose visual system is still functional (they are not blind). Hemineglect may apply to every range of space, that is, space of the body and around the body, proximal figural space, and large-size vista space. The patients do not detect the elements that are part of one half of their visual field. Only by insisting, may another person obtain that the patients eventually mention the elements that they spontaneously ignored. A dominant hypothesis refers to a deficit of attentional mechanisms and exploratory behavior normally triggered by the stimulations from the environment.

Besides the perceptual form of hemineglect, a representational form of it is also documented, in which people invited to evoke the visual image of an object or a scene leave the impression that they "ignore" the content of one of the hemifields (frequently, the left one) and only report or draw the elements present in the other (the right one). The most famous illustration of representational hemineglect is provided by the observations of two patients who suffered strokes that resulted in right parietal lesions, as reported by Edoardo Bisiach and Claudio Luzzatti (1978). The patients were Milanese residents. They were asked to imagine themselves standing on the Piazza del Duomo, facing the cathedral from a location at the opposite side of the square. They were then invited to describe every building, shop or other urban landmark visible from this place. The two patients showed the same bias, which consisted of naming essentially the buildings to the right of the visualized scene. When invited later to describe the square from the opposite point of view (with their back to the cathedral), the patients now described only the buildings to the right of the new perspective. This means that they described the elements that they had ignored and neglected those mentioned before (Figure 8.1). Thus, the whole spatial information is available, but the patients only access a portion of it, as defined by the perspective temporarily taken on the scene.[2]

The abundant literature on spatial neglect reveals the extent to which this pathology affects patients' daily life. One of the most notable consequences, as regards the processing of space, is the particular way patients imagine and describe routes. In particular, they refrain from describing paths which require turns to the left and prefer routes that only involve right turns. This strategy may even lead them to imagine improbable itineraries, longer and more complicated than those produced

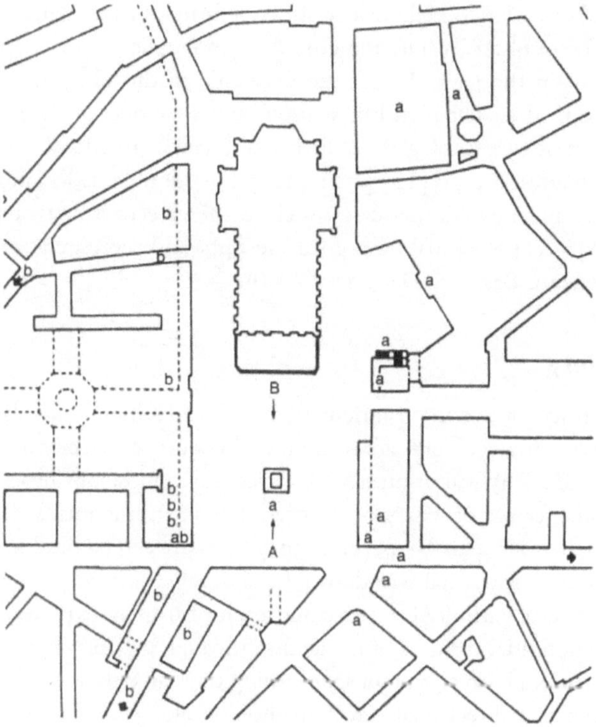

FIGURE 8.1 Map of the Piazza del Duomo in Milan. Letters A and B show the positions from which patients were asked to imagine the square. Letters a and b show the landmarks that patients recalled from each imagined position. Reprinted from the article by M. J. Farah (*Trends in Neurosciences*, 1989), by permission of the author and Elsevier

by a rational describer. A revealing situation is illustrated when, after describing a route from A to B, they are invited to describe the way back from B to A. Instead of generating a mirror description of the same path (with previous turns to the right becoming turns to the left), they prefer to describe quite another path, in such a way that ultimately the two routes considered altogether form a quadrangle. Such behavior suggests that route directions generated by these patients rely on mental images contaminated by the compelling representational bias that affects their whole cognitive life (see Bisiach, Brouchon, Poncet, & Rusconi, 1993).

Neuropsychology has documented a number of cases where both forms of neglect are associated in the same patient, thus suggesting the commonalities of the brain mechanisms that serve perception and mental imagery. But there are also cases of pure perceptual neglect, as well as cases of pure representational neglect. Remarkable cases of pure representational neglect have been described in the literature (see Beschin, Cocchini, Della Sala, & Logie, 1997; Guariglia, Padovani, Pantano, & Pizzamiglio, 1993), whereas a case of pure perceptual neglect particularly well documented is that of film-maker Federico Fellini, who suffered from a stroke which severely damaged his right hemisphere. While remembrance of familiar urban scenes by mental imagery was not in the least altered, the drawings produced by the artist showed some omissions in their left part, but without any alteration of their creativity and visual inventiveness (see Cantagallo & Della Sala, 1998). In addition to these clear-cut cases, which attest for a typical phenomenon of "double dissociation," there are also exceptional and bewildering cases, such as the case of a patient with bilateral lesions who exhibits perceptual neglect to the right and representational neglect to the left (see Beschin, Basso, & Della Sala, 2000).

Spatial anxiety

Space, as the place of accomplishment of human activities, may also be marked with uncertainty. Displacements are sometimes insecure, especially in environments perceived as unsafe. Physical or human obstacles may oppose our progression. Space is thus a potential source of worry or concern. In its extreme forms, anxiety engendered by an infelicitous spatial experience may be expressed by phobic reactions and even push people to irrational withdrawal behavior.

Upstream of such pathological reactions, each of us may experience discomfort because of the potential dangers of navigation through an unfamiliar environment. People who score high on an inventory assessing their anxiety level ("trait anxiety") make more frequent directional errors in moving along a route or drawing it as compared with people at the opposite end of the scale (see Viaud-Delmon, Berthoz, & Jouvent, 2002). On this matter, however, it is relevant to focus on specifically spatial forms of anxiety. A scale designed by Carol Lawton (1994) includes items that describe anxiety-provoking situations like: "Trying a new route that you think will be a shortcut without the benefit of a map," "Leaving a store that you have been to for the first time and deciding which way to turn to get to a destination," or "Finding your way to an appointment in an area of a city or town with which you are not familiar."

Measurements based on this scale reveal overall higher spatial anxiety level in women than in men. However, beyond gender-related differences, spatial anxiety is in general higher in people who claim to have difficulty in maintaining directional landmarks and cardinal directions in memory during their displacements. It is not a surprise that those people who hardly keep in memory their successive positions in an environment experience more anxiety when they consider the risk of losing their way. But the reciprocal line of reasoning makes sense as well, if one admits that anxiety generated by the fear of getting lost may lower people's capacity to focus on the cues that should help them to keep along the correct orientation (Lawton, 1994; Lawton & Kallai, 2002). Spatial anxiety is also expressed through the errors that affect directional pointing (Lawton, 1996) or when people must use directional instructions to control the moves of a toy car in a three-dimensional environment (Hund & Minarik, 2006).

In its most severe forms, spatial anxiety results in specific phobic syndromes. Agoraphobia is a pathological aggravation of the feeling of anxiety provoked by unfamiliar environments, whether large empty spaces or crowded places. For most people suffering agoraphobia, such environments generate panic and elicit avoidance behavior. The main psychiatric interpretation of agoraphobia is that this pathology results from the instatement of an enduring trouble following a particular experience of panic. However, more recently, agoraphobia tends to be considered as distinct from panic disorder, in spite of their frequent association. In any case, agoraphobia is typically linked with other expressions of anxiety.

Agoraphobia may occur in young people. It takes a particular form when it appears in elderly people. It is then associated with symptoms of depression, as well as other forms of spatial disorientation mentioned above in this chapter (see Ritchie, Norton, Mann, Carrière, & Ancelin, 2013).

Notes

1 Postmortem examination of the patient's brain in fact revealed that a large portion of the hippocampus had been preserved after the neurosurgical procedure that 27-year-old HM underwent, contrary to what MRI scans indicated forty years later (see Annese et al., 2014).

2 Twenty-five years prior to Bisiach and Luzzatti's (1978) article, to which hundreds of scholars have referred and continue to do so, a clinical observation of similar meaning was reported by Macdonald Critchley, a famous British neurologist, in his book *The Parietal Lobes* (1953). The patient described by Critchley showed a right parietal lesion. A task consisted of drawing a sketch map of Piccadilly Circus (with north at the top of the sheet of paper, as usual). Several streets and avenues converge to the famous London square from different directions. It was striking to see that the patient did include all those ways on the map, but drew all of them as connected to the right half of the perimeter of the square.

9

SPATIAL CHALLENGES

Navigating is always a challenge. Admittedly, most of our journeys unfold without any serious risk or uncertainty. When we say that we follow a route "with our eyes closed," the phrase makes it explicit that the locomotor program implemented does not involve any conscious processing of landmarks and directions. Vision is essentially engaged for controlling the alignment of trajectories and detecting obstacles.

Actually, there are situations that give us the opportunity of experiencing some kind of adversity while we are progressing, and for which friendly assistance is welcome.

Cognitive and locomotor impairments

A number of neurological diseases are likely to impact cognitive functions and, more specifically, locomotion and spatial behavior. Young adults with intellectual disability are typically confronted with arduous situations when they try to achieve autonomous mobility in their environment. For instance, in learning new routes, they need help in identifying the most relevant and distinctive landmarks at reorientation points (see Courbois, Blades, Farran, & Sockeel, 2013). Another challenge consists for the relatives of these patients (adults with intellectual disability or children with Down syndrome) to assist them in enhancing capacities for spatial reasoning, such as those needed to find shortcuts (see Courbois, Farran, Lemahieu, Blades, Mengue-Topio, & Sockeel, 2013; Mengue-Topio, Courbois, Farran, & Sockeel, 2011).

Motor disorders are particularly noticeable in people suffering from Parkinson's disease (PD), which is also known to be expressed through various forms of cognitive impairment. The cognitive dimension is well demonstrated in studies where motor load is removed by the experimenter, for instance in the case of traveling by car. Studies have investigated people diagnosed as suffering a moderate form of PD that did not prevent them from driving. Compared with people without neurological disease, PD patients who receive verbal instructions to move along an urban route make more errors at reorientation points and are more likely to get misled off the route. They furthermore make more frequent driving errors with potential impact on security. They also perform less well in visual and cognitive tests. In a situation where attention is in heavy demand, the capacity limitations of the executive system are manifest (see Uc, Rizzo, Anderson, Sparks, Rodnitzky, & Dawson, 2007).

Other motor challenges are met by people who are affected by cerebral palsy (CP). Most of the patients who obviously need assistance in motion in fact have no trouble as regards route planning. It remains, however, that the motor limitations of the patients, with marked hypertonia and spasticity, massively impact their displacements. Their motor difficulties are responsible for restriction, and even deprivation, of sensory experiences associated with moves. The costly control of motricity calls for extra cognitive resources and competes with the attention devoted to the environment. A feeling of insecurity may also make CP patients' life particularly difficult. Lastly, the arduousness of physical activity implied by any travel results in a vicious circle, with worsened locomotion and fewer opportunities for spatial experience (see Opheim, Jahnsen, Olsson, & Stanghelle, 2009).

Another neuromuscular pathology, highly invalidating although not associated with cerebral damage, is spinal muscular atrophy (SMA). Motor disease in children suffering from this illness is such that no motion is ever possible for them. However, children do not display any cognitive deficit in the spatial domain, thus providing indirect support to the theory that confers predominant function to visual perception in coding spatial information (contrary to sensorimotor theories in line with Piaget's views). The illness is not associated with any delay in the understanding and use of spatial expressions. The good control over spatial language on the part of these children is ascribed to the fact that due to their absolute deprivation of motor experience, they more systematically rely on language to identify objects, their positions, and the actions they request their relatives to execute. The positive side of the situation of these children is that in the absence of locomotor experience, their spatial cognition seems to develop in a quasi-normal fashion (see Rivière & Lécuyer, 2002; Rivière, Lécuyer, & Hickmann, 2009).

Space without vision

There is no doubt that the continuous stimulations conveyed by vision contribute to substantiating our experience of space. However, such experience is also possible, at least within certain limits, in the absence of vision. When such is the case, the issue is to assess the extent to which a person's locomotor performance is affected. Would spatial orientation and autonomous navigation suffer from this particular form of sensory deprivation?

Limitation of performance of blind people in complex environments is attested by a number of experimental observations. For instance, when blind people reconstruct a familiar route in the form of a map model, they posit landmarks along the route as correctly as sighted people do, but they experience more difficulty with landmarks that stand off the route and whose localization requires relying on a survey representation of the environment (Hollyfield & Foulke, 1983). It has also been shown that congenitally blind people have more difficulty than adventitiously blind and sighted people when they are asked to point to objects in an environment that they have learned by locomotion (Rieser, Guth, & Hill, 1986). In the same vein, compared with late blind and sighted people who learned a route while blindfolded, the congenitally blind make greater errors in their judgments of direction and distance (Veraart & Wanet-Defalque, 1987).

Further works have confirmed such differences, although of much lower magnitude, suggesting that the former claims need to be qualified. This was made in particular by Jack Loomis and his colleagues by taking up the experimental situation cited above (Rieser et al., 1986). In distance estimation, only small and non-significant differences were in fact found between performances of groups with different visual experience (Loomis et al., 1993). The same effect was also demonstrated in a study on estimation of absolute distances between objects in an environment learned by guided locomotion. The study provided findings that did not suggest any qualitative differences between the early blind, the late blind, and the (non-blindfolded) sighted. Visual experience thus does not seem to be essential for elaborating and using consistent spatial representations that reflect veridically an environment's metric (see Haber, Haber, Levin, & Hollyfield, 1993).

These findings attest for the proficiency of blind people, including early blind, to control spatial behavior and the building of representations that support such behavior. They force researchers to question the commonly recognized role of vision in spatial knowledge. Of course, other sensory sources (auditory, kinesthetic, haptic) are acknowledged as providing cues to build up representations of spatial environments. But we cannot pass over the widespread intuition that vision might be the "spatial sense" *par excellence*. This intuition has been questioned in a thorough review (see Thinus-Blanc & Gaunet, 1997), as well as a more recent analysis by Zaira Cattaneo and Tomaso Vecchi in their book, *Blind Vision* (2011).

An analysis of the issue requires reviewing the purported advantages of vision over the other sensory modalities as regards space processing. Frequently mentioned is the inherent capacity of the visual modality to operate *simultaneous* processing of information, in particular large volumes of such information. Other senses appear to be more specialized in temporal processing of information and in capturing its sequential features. This peculiarity of non-visual modalities confers them precedence in the construction of sequential spatial representations. It might account for the fact that the early blind are not disadvantaged in the encoding of routes, while they experience more difficulty with the building of allocentric representations, which are easier to construct from visual experience. It is well established that they are less proficient than sighted people at using the allocentric perspective.

It is relevant here to distinguish between two families of processes: first, those controlling *memory* of experienced space, for instance when a task consists of reproducing a previously learned route or drawing a route map showing landmarks along the path; and second, those involved in the production of *inferences*, which call for various forms of reasoning, such as computing the relative positions of several landmarks or making directional judgments from various vantage points. For the early blind, who do not perform lower than the sighted in spatial memory tasks, the greatest difficulty is with tasks that require inferences (see Thinus-Blanc & Gaunet, 1997).

These conclusions would be quite acceptable if the literature did not provide examples revealing a still more complex situation, not only with attestations of the absence of difference between the sighted and the blind, but even with case reports of blind people's higher performances. Carla Tinti and her colleagues investigated whether inferential capacities of the early blind (which are supposed to rely on allocentric representations of the environment) were really lower than those of the sighted. Their participants learned to move along two pathways (10 and 15 meters) with several right-angle turns and designated landmarks. The participants moved autonomously along the pathways, without any assistance or guidance. Measures were then taken of their locomotor performance (pathway completion, with count of errors), direction and distance estimates to landmarks from different points of the pathways, and hand-drawn reproduction of the pathways. The results showed higher performance of the early blind in all tasks, including those calling for allocentric representation (see Tinti, Adenzato, Tamietto, & Cornoldi, 2006).

The objective of studies in this domain is not only to test *navigational capacities* of non-sighted people, but also to capture some reflection of the *representations* built by these people from locomotor episodes. In a spatial learning task to be reported now, early blind, late blind, and blindfolded sighted participants were invited to walk around freely in a room of 24 square meters and learn the positions of six virtual objects identified by a localized sound source (a telephone, a clock, etc.). Learning was considered to be completed when the participants, standing at the center of the room, were able to point correctly to the location of each object (with the sound sources being turned off). The structural properties of the mental representation of the environment were then tested by asking the participants to imagine pairs of objects and mentally scan the distances separating them. In blind participants, mental scanning times were strongly correlated with actual distances experienced during learning, suggesting that their internal representation did possess metric properties reflecting the structure of the environment. But although the sighted participants had met the same learning criterion as the blind, their mental scanning times were completely uncorrelated to distances, which indicated that they were unable to build a metrically valid representation of the environment they had explored while blindfolded (Figure 9.1). This is an example of the capacity of blind people to build genuinely spatial representations embodying veridical metric information (see Afonso, Blum, Katz, Tarroux, Borst, & Denis, 2010).

Further attestation of the remarkable navigational capacities of blind people may be found in other contexts of acquisition of spatial information. This is the case

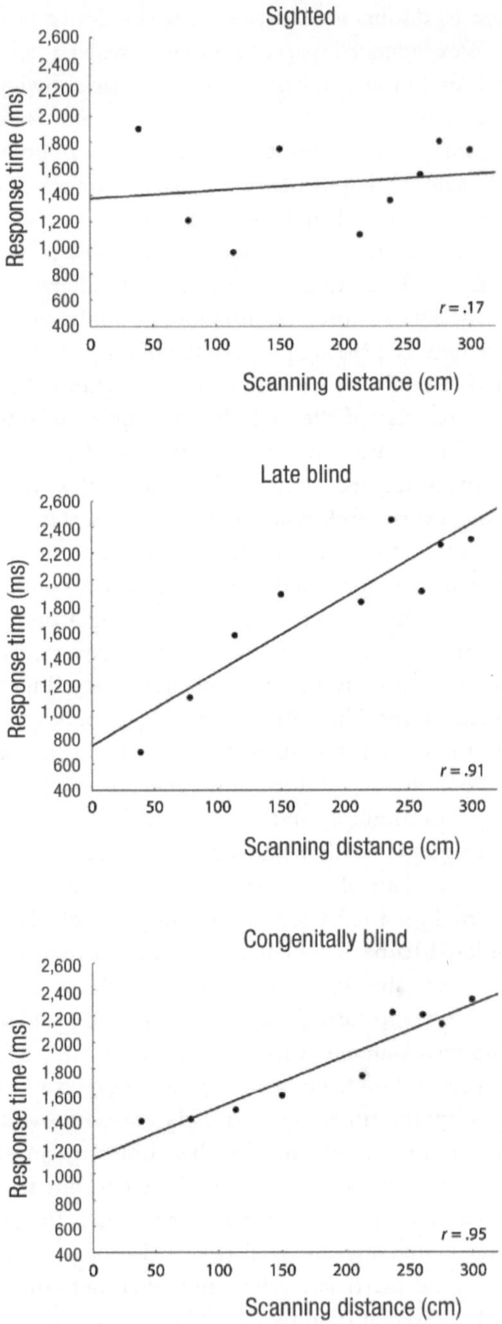

FIGURE 9.1 Mental scanning time as a function of distance after spatial learning by sighted, late blind, and congenitally blind participants. Reprinted from the article by A. Afonso, A. Blum, B. F. G. Katz, P. Tarroux, G. Borst, and M. Denis (*Memory and Cognition*, 2010), by permission of Springer

when the learning of a new environment takes place off-site, outside of any actual navigational episode, with the help of a virtual device based on audio information. In a follow-up study to the experiment reported above, blind people were invited to learn an environment in a research center. The environment consisted of corridors along which offices were distributed on both sides. Particular sounds were emanating from some of these offices (keyboard clicking, printer, electronic music, conversation, etc.).

A first group of blind participants was invited to freely explore this environment. They were offered the possibility of going back and forth along the paths as many times as desired, until they were sure to have understood the spatial configuration of the environment. They later completed tests intended to assess the quality of the representation they had constructed of the environment. For another group, participants were presented with an audio virtual reality device simulating the same environment. The participants were sitting on a chair and generated their moves along the corridors by using a joystick, having every possibility of moving back and forth. The device provided them with the acoustic cues associated with each of the offices. The participants were equipped with an orientation sensor which made it possible for the sounds to be veridically affected by every head reorientation. To summarize, in the absence of any locomotor experience of the environment, the participants were invited to become familiar with its spatial structure by using the acoustic cues provided by the virtual device.

The results of the tests showed that the participants who experienced the virtual device performed similarly to those who actually walked in the environment. In particular, they were quite proficient at mentally comparing distances between pairs of places associated with specific sounds. They were also able to correctly reconstruct the navigated environment by using Lego® bricks. The study thus attested blind people's capacity to acquire consistent, valid configurational knowledge of a new environment from a set of spatialized sounds. As is found with actual navigation, virtual navigation makes it possible to create coherent mental representations that preserve topological and metric properties of an environment, a characteristic that confers on such type of representation the status of a "cognitive map" (see Picinali, Afonso, Denis, & Katz, 2014).

Obviously, a total lack of visual experience since birth does not preclude the setting up of an effective spatial representational system. It remains that if visual experience is of major assistance for building an allocentric representational system, people deprived of such experience nevertheless succeed in doing so, although at greater cognitive expense (Pasqualotto, Spiller, Jansari, & Proulx, 2013). We should also note that using an allocentric reference frame is easier in tasks that apply to near space rather than large-size spaces inviting to locomotion (Iachini, Ruggiero, & Ruotolo, 2014).

By privileging comparisons between groups of people with different visual histories, researchers sometimes overlook an important aspect of the issue, namely, the existence of interindividual differences *within* groups of blind people. The evidence of such differences in locomotor tasks is frequently sought between the congenitally and the adventitiously blind, who are supposed not to have the same capacities, due to their respective visual histories. It is acknowledged today that the deciding factor of locomotor performance is not the age of onset of blindness, but rather people's quality of motor experience and level of autonomy. It is well known that particular learning experiences, as well as factors related to the social environment, predispose individuals to reach different levels of autonomous mobility (see Loomis et al., 1993). Note also that impairments observed in early blind people must not be interpreted as reflecting mere absence or unavailability of the processes that are implemented by sighted people. Difficulties weighing on the acquisition of efficient processing capacities must be considered, instead of sticking to the presumed absence of these capacities. Many factors may hinder spatial behavior of non-sighted people, although some of them aptly implement compensatory strategies. In some cases, the development of such compensatory processes is costly and may result in lower performance (see Cattaneo et al., 2008).

The nature of spatial representations

Blindness raises well-founded questioning about its impact on spatial knowledge, but at the same time opens to a set of more basic questions, such as that of the nature of spatial representations, not only those built by people deprived of the visual sense, but also by sighted people. Beyond the representations of space derived from various sensory modalities, shouldn't we envisage that the cognitive system builds and entertains spatial representations that are neither visual, nor kinesthetic, nor even linguistic, but in fact untied to any particular modality under which information has come to a person's knowledge?

One of the first formulations of this hypothesis came from a study by Carreiras and Codina (1992), in which the experimenters submitted early blind and sighted participants (blindfolded or not) to the learning of 31 × 24-centimeter labyrinth-type scale models representing fictitious cities, including a network of streets and relief blocks of buildings, as well as landmarks (drawing pins of different textures and colors). Later, the participants' task consisted in estimating route and straight-line distances between all possible pairs of landmarks. The results showed that performance was quite similar for the three groups of participants (except as regards response times, overall shorter for the non-blindfolded sighted). Based on these findings, Carreiras and Codina developed the idea that the representation of spatial information does not necessarily rely on *visual* representations (or representations from any other sensory modality), but on representations that they considered to be *amodal*.

Parallel to studies specifically devoted to blind people, this idea was promoted in the form of the claim that spatial representations derived from various modalities (vision, audition, touch, language, etc.) are functionally equivalent, or possibly identical, and that they converge on a common representational system where they lose their original sensory mark (see Avraamides, Loomis, Klatzky, & Golledge, 2004; Bryant, 1997; Loomis, Lippa, Klatzky, & Golledge, 2002).

The idea was taken over in the context of studies on spatial cognition of the blind, with the postulate that good performance of the blind in this domain cannot be attributed to visual function. The works elicited by this idea went beyond the domain of navigation and considered also the cognitive representation of configurations and spatial scenes. When such representations are postulated in blind people and of course cannot be said to have a visual substrate, the question of their nature is raised. This theoretical question, then, goes much beyond that of blind people's spatial representations, although having originated from this particular domain of research.

Marijn Struiksma and her colleagues from the universities of Utrecht and Twente provided arguments for the distinction between the notion of *visual image* (a mental representation derived from a particular sensory source) and that of *spatial image*. The latter, rather than being simply qualified as amodal, is claimed to be *supramodal*, by the fact that it integrates pieces of information coming from all the modalities contributing to a particular experience, without however the reference to these modalities being lost. But a spatial image is in no way reducible to a plain visual image. The concept of spatial image is relevant to account for the functional equivalence of spatial knowledge issued from perceptual entries and language. As a matter of fact, the approach takes language into account as one of the means available for constructing spatial images, an idea that has been illustrated for a long time in research on mental imagery (see Denis & Cocude, 1992; Kosslyn, Reiser, Farah, & Fliegel, 1983; see also chapters in Part IV). On the whole, the low differences observed between performances of blind and sighted people are thought to reflect the similar involvement of this cognitive function (and the associated brain structures) for both groups (see Struiksma, Noordzij, & Postma, 2009).

In recent years, fMRI studies have made it possible to identify the brain correlate of supramodal representations, namely, the left supramarginal gyrus. To be functional, this structure does not require visual experience to develop. Its activation is found in both blind and sighted people invited to judge spatial statements, a finding supporting the view that supramodal representations are likely to be generated from spatial language. Besides, during processing of spatial language, the blind and the sighted activate quite similar brain networks. The only differences are those resulting from the reorganization of the occipital cortex in blind people (see Struiksma, Noordzij, Neggers, Bosker, & Postma, 2011). In any case, visual experience is definitely not a prerequisite for the development of supramodal representations of spatial information (see Ricciardi, Bonino, Pellegrini, & Pietrini, 2014; Ricciardi & Pietrini, 2011).

In fact, although this conception has not received full agreement today (see Pasqualotto & Proulx, 2012), it has been taken over and aptly formulated by Loomis, Klatzky, and Giudice (2013): "Spatial images [which are conceived by these authors as amodal] can integrate inputs from multiple perceptual and cognitive channels; they transcend modality-specific features while retaining the spatial information shared by the inputs" (p. 134).

Navigational assistance to the blind

Assistance to the navigation of blind pedestrians is an important endeavor for the acquisition or maintenance of their autonomy in locomotor space. The issue has motivated a number of research programs based on more or less sophisticated technologies. Navigational aids may be delivered at different steps: preliminary planning, navigation along the selected path, and control of progression (avoiding unforeseen obstacles). At each of these steps, specific technological resources are required. Of course, their implementation requires taking into account users' expectations and preferences (see Gallay, Denis, & Auvray, 2013).

A well-assessed system of this kind is the Personal Guidance System that was developed by a research team from Santa Barbara. The system includes a Global Positioning System (GPS) module that locates the navigating person (thanks to an electronic compass), a GIS that delivers directional information, and an interface that transmits guidance information to the user through headphones. The system emits spatialized beeps which enable the blind person to get oriented in the proper direction. The use of tones is frequently perceived as more appropriate than a system delivering verbal instructions (for instance, "to the left, 80 degrees"), which is costlier in terms of cognitive load. The advantage of the present system is to convey a signal that can be directly interpreted, since it is geometrically matched to the spatial directions to which it refers. A variant of the system uses a pointer that the navigating person orients in what he/she thinks is the right direction. The system reacts by issuing a sequence of beeps whose intensity increases as the point to reach is approached. This device based on spatialized tones is appreciated by its users, even if the dependence on continuously emitted auditory information through headphones is sometimes experienced as a limitation of the system (see Loomis, Marston, Golledge, & Klatzky, 2005; Marston, Loomis, Klatzky, Golledge, & Smith, 2006). Lastly, in experiments where cognitive load is imposed on travelers during navigation (for instance, by imposing a secondary task that weighs on working memory), performance is better when users receive directional tones rather than verbal directions (see Klatzky, Marston, Giudice, Golledge, & Loomis, 2006).

Other systems are based on the tactile sense, by using a vibrotactile stimulator worn inside a wristband or attached to the torso. The signal delivered through such systems ensures a function that is similar to that of spatialized beeps, by providing information on the direction to take and the heading to maintain along the route. This device can be extended to other situations involving sighted people, such as the displacements of emergency personnel in environments without visibility (see Marston, Loomis, Klatzky, & Golledge, 2007). A similar system has been developed by a multidisciplinary team from Oldenburg, Germany. Tactile Wayfinder uses a navigation module connected to a belt with an electronic compass and six vibrators that indicate directions and deviations from the path. Direction information is thus immediately intelligible. In an advanced version, the system not only delivers the directional information about the next waypoint to reach, but provides advance information about the next two waypoints, in such a way that the user can anticipate how the route continues when reaching any point (see Heuten, Henze, Boll, & Pielot, 2008; Pielot & Boll, 2010).

Research on assistive devices for blind people has also inspired systems whose benefit extends to other users, while remaining fully appropriate to the originally targeted ones. As an illustration of this approach, the same German team has designed PocketNavigator by installing a tactile interface in a smartphone. Each of the eight directions (ahead, left, right, behind, and the four intermediate directions) is associated with a specific vibration pattern. For instance, in order to indicate that a user has to walk towards the next waypoint, two short vibration buzzes indicate "ahead." When the next waypoint is to the right-hand side, a short buzz is followed by a long one. Directions are given in relation to the user's walking direction, which is obtained from the GPS signal. The advantage of the system is that the device may remain in the user's pocket, which does not constrain his/her gestures. Furthermore, eyes-free usage avoids distracting the user from the environment. The disadvantages are that the vibration patterns have to be learned and that walking directions cannot be determined when the user does not move (see Pielot, Poppinga, Heuten, & Boll, 2012).

With such devices, distal information on the environment is not exploited, which has the effect of sticking the non-sighted walker to egocentric strategies, at the expense of the allocentric perspective, much more easily adopted by sighted people. However, despite these differences, the spatial images built by the sighted and the visually impaired are thought to share similar metric properties (Noordzij, Zuidhoek, & Postma, 2006). The question remains whether the assistive devices based on GPS systems entertain users' "dependence" on the egocentric perspective. This perspective is highly available to blind people, although they are not prevented from building allocentric representations when this is valuable. The security ensured by passively keeping up with instructions that come from a satellite service considered as reliable nevertheless hinders the emergence of alternate navigational strategies.

PART IV
Space and language

Until now, we have focused on space as it is apprehended by the senses and experienced through locomotion. As cognitive scientists, we have been interested in the mechanisms by which space and navigational behavior generate internal representations on which human cognition performs computations and plans travels.

Part IV is concerned with space when it is subjected to *symbolic mediation*. In other words, we now take into account systems interposed between the senses and action. While the mechanisms discussed above apply to the whole range of living creatures with locomotor capacities, the issues to be addressed here are restricted to the species endowed with the most extended and diverse set of symbolic functions – human beings.

Dealing with symbolic mediation processes implies considering the properties of symbols used to carry the representational function attached to them. Each form of representation imposes particular constraints (and brings specific benefits) to cognitive agents. In this part, we will devote special attention to language, a symbolic system whose internal organization and rules of functioning differ from those of objects to which it applies. Language is a sequential mode of communication made up of discrete units distributed according to an arbitrary organization. In sharp contrast, space displays all of its content simultaneously, its structure is continuous and the representations constructed from it show a high degree of analogy with respect to represented entities.

We then face a dual challenge. One consists in translating spatial experience into linguistic expressions, while the other is the reconstruction of that experience by the recipient of the message. Completing these operations entails interfacing the modules in charge of processing information issued from the two systems. The cognitive architecture is not only endowed with functionalities specific to the two systems of representation, but also with mechanisms that translate the content of one system into a format that is readable by the other (see Bloom, Peterson, Nadel, & Garrett, 1996).

10

SPATIAL TERMINOLOGY

Language is one of the means available to convey spatial knowledge or report a travel experience. It enables us to deliver to someone the description of a scene to which the person does not have, did not have, and may well never have any perceptual access. A possible substitute for direct perceptual experience is to use a recording system capable of transmitting a veridical and trustworthy sensory equivalent of the scene (photograph, video, hologram). When there is no way to generate such a document, language is available as a tool to help an addressee build the spatial representation of a distant scene. To be successful, communication requires that the speaker disposes of adequate linguistic and expressive capacities. It supposes also that the addressee has the capacities needed to build a representation that validly encodes the description. Coordination of both sets of competences is thus required, which supposes that partners share the same linguistic references and cognitive schemas for the representation of space. This is, in particular, essential in the context of conversation (see Rodrigo, de Vega, & Padrón, 2012; Schober, 1993, 2009).

Spatial language

Spatial language takes many forms and serves varied objectives: to describe a scene, to refer to the place occupied by an object, to provide route directions, etc. In its elementary form, the case considered here is that of the relationships between a (real or imagined) scene involving at least two objects and a statement describing their relative positions. It is the function of spatial prepositions to denote such relations (containment, support, contiguity, etc.). For instance, "The key is on the table." The relation may concern more than two objects: "The Picasso is hung between the Miró and the Dalí." Descriptions may also apply to events ("The dog passes behind the armchair").

A seminal contribution to the comprehension of spatial expressions is due to Barbara Landau and Ray Jackendoff (1993). By referring to "the language of objects and places," these authors pointed to the contrast between two basic questions about an environment. *What* object is there? *Where* is it located? Indeed, "what" and "where" are the two facets of every spatial situation, which have become familiar to us since neurobiology has described the brain circuits activated by these representations (see Ungerleider & Mishkin, 1982). Since this major discovery, neuropsychology has explored the two functions and evidenced the specific deficits that affect them. In a comprehensive theoretical frame, Landau and Jackendoff accounted for the mechanisms by which language refers to objects and expresses their spatial relations. Their approach had the value of showing that the analysis of spatial language not only informs us on language, but also on the mental representation of space.

The main function of spatial prepositions resides in their capacity to specify topological relations between objects. According to a classic distinction, the speaker refers to two objects that are assigned the roles of "figure" (the located object) and "ground" (the reference object), respectively. Their relation is expressed by a preposition which specifies the portion of space where the figure-object is located relative to the reference-object.

With language, however, the dimensionality of objects and situations is not communicated in an analog manner. While the size of objects varies along continuous dimensions, the terms used cut out these dimensions in a binary fashion ("tall"/"small," "thick"/"thin," etc.). Spatial language, in its natural use, is not prepared to convey fine metric information. If there is absolute need to communicate such information, speakers will commit themselves to formulations that include measurement units. As a consequence, the formulation may reach an unnecessarily high degree of metric specification, thus involving costly cognitive processing. For instance: "The vase is on the shelf, in the right angle, 10 centimeters from the edge of the right side, 15 centimeters from the front edge, and 5 centimeters from the wall behind."

The example is borrowed from Michel Aurnague, Laure Vieu, and Andrée Borillo (1997, p. 73), who developed special interest for the imprecision of natural language applied to the description of spatial relations. Some expressions are vague and unclear ("to be beside"). A statement about an object that is "in front of" another one does not say anything about the distance between them, neither whether the figure-object is slightly offset to the frontal axis of the reference-object. However, if we have knowledge about the particular objects concerned, a measurement scale is indirectly suggested. If we listen to the sentence: "The glass is in front of the bottle," we can imagine that the distance is in the order of centimeters, whereas for "The car is in front of the building," it will be evaluated in meters or tens of meters. This strategy of contextual relativization is very useful, since it helps to reduce the imprecision of an expression by making use of pragmatic knowledge. Aurnague and his colleagues insisted on the value represented by such moderated precision. In many contexts, applying the Gricean maxim of quantity is relevant:

"Do not make your contribution more informative than is required" (Grice, 1975). Too much specification is not only unnecessary, but is also counterproductive for communication.

Geometric notions present in spatial terminology, as they are carried by spatial prepositions, do not make it possible to fully represent the semantics of space. The context of enunciation must obviously be taken into account.

The language of spatial prepositions

Cognitive approaches to spatial language are reliant on the linguistic analysis of spatial prepositions. Two major contributions have marked the field.

The first one is due to Annette Herskovits (1986). According to her, literal interpretation of prepositions is not adequate and we must take into account the context and the conditions of use of language to account for their polysemy. For each spatial preposition, Herskovits identifies an "ideal" meaning, which corresponds to the geometric (prototypical) idea attached to it. But the actual use of a preposition is determined by semantic shifts that reflect some "admissible tolerance" for deviations from the prototypical meaning. For instance, the preposition "on" conveys both meanings of support and contiguity (between the supporting object and the supported one), which are both satisfied by a sentence referring to "the atlas on the table." Note that the sentence is still acceptable if a tablecloth or another book lies between the atlas and the table (preventing direct contact between them) (Figure 10.1). A shift of meaning of the preposition "on" in a sentence referring to "the wrinkles on someone's forehead" is another example of admissible approximation.

Similarly, while the preposition "at" is used to indicate the spatial coincidence between two entities ("The center of a rectangle is at the intersection of the two

FIGURE 10.1 "The atlas is on the table"

diagonals"), tolerance is accepted for a sentence like "Mary is at the gate." Mary and the gate obviously are not coincident. They are only very close together. The shift that is likely to occur from the geometric meaning of a preposition is gradual, with a threshold beyond which it is no longer appropriate to use that preposition.

Another classic example is provided by the preposition "in" and the various geometric descriptions corresponding to it. In the case of "the water in the vase," the liquid actually fills the "volume of containment" of the vase. But accepting a statement about "the bird in the tree" implies that the bird is somewhere in the interior of the three-dimensional volume of the upper portion of the tree (branches and leaves).

Language is tightly dependent on context. At the time of the founding analysis proposed by Herskovits, models of artificial intelligence for the processing of language were emerging, but these aspects were not yet really taken into account.

The other main contribution to the domain is by Claude Vandeloise (1991), the most insightful analyst of spatial prepositions in French. His basic premise is the same: the analysis of spatial markers of language helps the researcher to evidence the properties of speakers' mental representations of space. Beyond plain logic description of spatial prepositions, the intent is to develop a functional description which relies on extra-linguistic knowledge shared by speakers of any given language (including knowledge pertaining to naïve physics). Among these pieces of knowledge, the anthropomorphic dimension is given special consideration, by taking into account the privileged directions that derive from the structure of the human body (such as its symmetry). Vandeloise promotes a view of linguistics anchored in a cognitive approach and forerunning what will emerge later as the theory of "embodied cognition."

Vandeloise focuses on the asymmetry between the figure-object and the reference-object (in his terms, the "target" and the "site"). In the majority of cases, the expression of spatial relations is asymmetrical. If everyone clearly understands the sentence stating that "the book is on the table," very few would be ready to consider that "the table is under the book." In the acceptable sentence, attention is focused on the relation of support. It is also obvious that objects differ in terms of size or mobility. The object that is bigger and/or the most stable is typically coded as the site relative to which the target is posited. Even symmetrical relations, such as neighborhood ("to be close to"), are subordinated to this rule. While everyone easily understands the relations expressed by "The bicycle is close to the house," the mirror sentence ("The house is close to the bicycle") is beyond the limits of acceptability.

Vandeloise was also committed to analyzing the ambiguity of prepositional expressions such as "in front of" and "behind" (Figure 10.2). The sentence "The armchair is in front of the chair" correctly describes Scene 1, just as it also does for Scene 2, even though the position of the pieces of furniture is inverted. The use of "in front of" to describe the first scene is imposed by the orientation of the chair and is not affected by the presence of an observer. On the contrary, when the same sentence describes the other scene, its adequacy fully depends on the presence and position occupied by the observer/speaker.

FIGURE 10.2 "The armchair is in front of the chair." Reprinted from the book by C. Vandeloise, *L'Espace en Français: Sémantique des Prépositions Spatiales* (1986), by permission of Editions du Seuil

Vandeloise's book offers a systematic review of the use of spatial prepositions and expressions, including conventional formulations, such as the fact that in French as well as English, a person is said to be "sitting *on* a chair," but "sitting *in* an armchair." The accentuated concavity of a seat seems to conceptually change the perception of a relation support/supported into that of container/contained. The analyses confirm the idea that geometry and logic are unable to satisfactorily describe the rules of usage of spatial prepositions. A consideration of world knowledge and pragmatics is required to account for spatial language.

Linguistic diversity in the use of prepositions

In any given linguistic area, spatial language reveals its ambiguities and peculiarities. But the comparison between languages also reveals how diverse they are in regulating the use of spatial prepositions.

We will focus here on two illustrative contributions. The first one comes from Melissa Bowerman (1996), who showed from comparing languages that the analysis of spatial scenes and their verbalization may not involve the same semantic categories and thus results in different uses of spatial prepositions. Consider situations involving relations of support, contact, and containment, as expressed in four languages (Figure 10.3). In many languages (including English and French), the relations involving contact with and support by a vertical surface (here, "the handle *on* the door") are treated as similar to relations of contact with and support by a horizontal surface ("the cup *on* the table"). But a distinct preposition is used to refer to the relation of containment ("the apple *in* the bowl"). The groupings are quite different in the case of Finnish. And the situation is still different for Dutch, which uses three distinct prepositions, while Spanish does not differentiate at all among the three cases.

Claude Vandeloise (2008) proposed a thorough analysis of three prepositions in two languages. English prepositions "at," "on," and "in" (and their French counterparts "à," "sur," and "dans") are among the most frequently used spatial prepositions, but also, mainly as regards "on" and "in," the very first acquired by young children.

Consider two sentences: "The child is in Rome," "The child goes to Rome." English uses "in" for designating the current position of a target, but uses a distinct

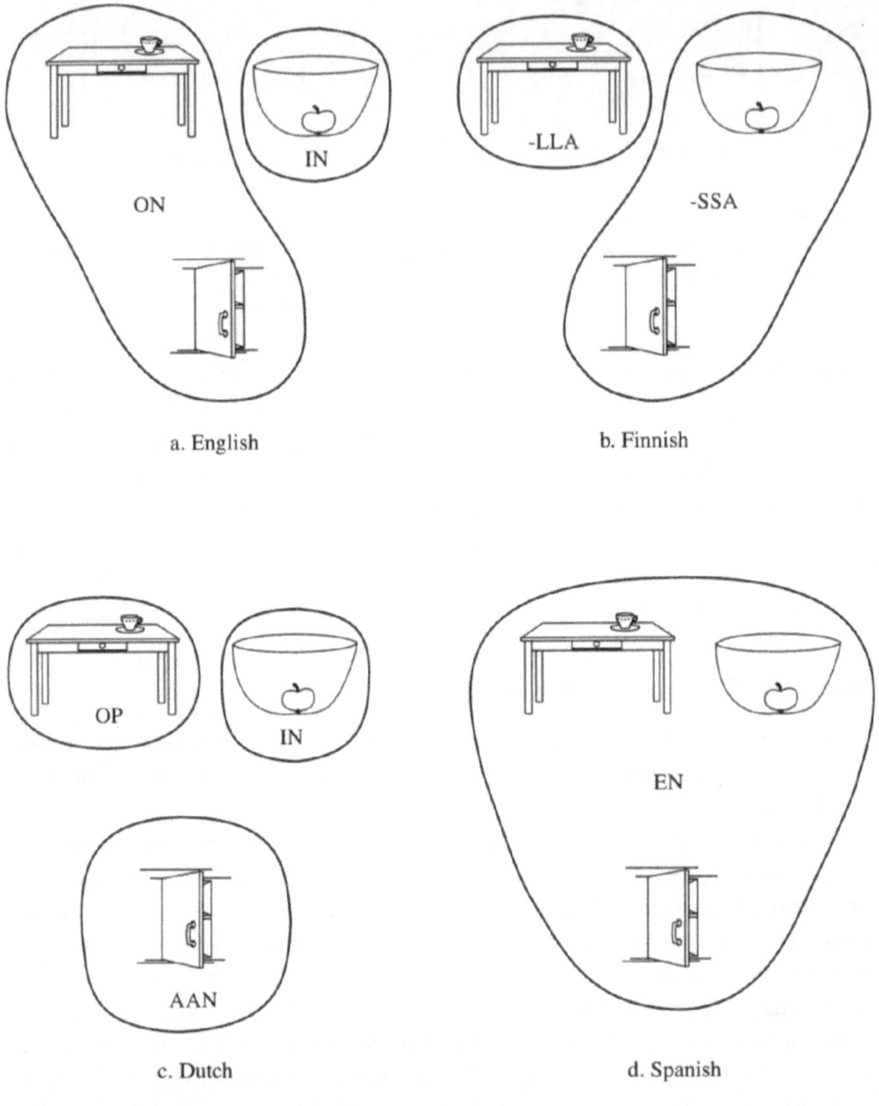

FIGURE 10.3 Classification of three spatial situations in English, Finnish, Dutch, and Spanish. Reprinted from the chapter by M. Bowerman in *Language and Space* (1986), by permission of The MIT Press

preposition ("to") to indicate its prospective position. French uses the same preposition "à" to refer to either current or prospective positions ("L'enfant est à Rome," "L'enfant va à Rome"). Obviously, in the two linguistic systems, these locative prepositions have different extensions. English has other prepositions for referring to the prospective position of a mobile target. In addition to "to," it can use "into" and "onto" ("The child is walking to the theater," "The child goes into the house," "The cat is jumping onto the table"). French does not make it possible to introduce the prospective position of a target with a verb that would convey the manner of displacement (here, "marcher") followed by a preposition (here, "à"). The best approximation in this case would be to use prepositions "vers" or "jusqu'à" ("L'enfant marche vers le cinéma," "L'enfant marche jusqu'au cinéma"). Another solution might consist in stating: "L'enfant va au cinéma en marchant" (which in fact does not sound natural). Here appears the value for English to have "to," "into," and "onto" at its disposal. In English, the verb expresses the manner of displacement and the "satellite" particles convey the expression of direction ("up," "off," etc.), whereas in French, the verb expresses the direction, and the mode of displacement is not necessarily specified or is optionally conveyed by a complementary expression ("en marchant," "en courant," etc.).

The prototypical spatial uses of English prepositions "at," "in," and "on" do correspond to those of the French prepositions "à," "dans," and "sur." But the expressions validated in both languages are not the same. In particular, there are more occurrences of the use of "in" to introduce spatial entities than is the case for "dans." English uses "in" for names of countries and cities, reflecting that these entities are treated as containers, whereas French, which uses "à" for city names, apparently treats cities as locations. To locate and to contain correspond to functions that are not considered the same way in both languages.

Spatial perspective and frames of reference

When visiting an environment, observers have a perspective, and only one, on the scene available to their sight, as determined by their currently occupied position and viewing orientation. Of course, observers may well escape this particular perspective, for instance when being instructed to imagine the scene as it would appear if they occupied the place of a designated object and were oriented towards another. Observers may also be invited to imagine the point of view of other observers present in the environment. Thanks to language, they are able to describe alternate points of view by imagining other frames of reference than the one defined by their position in space. As a result, the descriptive statements will differ, although the same set of operators will be used (the same repertoire of spatial prepositions), in order to characterize the relative positions of objects from decentered perspectives. In any case, speakers cannot elude the obligation of adopting a particular frame of reference and making their best to have the addressee adopt it. But on what exactly should the frame of reference be centered?

The frame of reference centered on an observer is said to be *deictic*. When an object is declared to be "to the left" or "to the right" of another one, it must be understood that what is referred to is the left or the right of the observer, as the observer is presently oriented when producing the statement. A third party will comprehend such a spatial description only on the basis of knowledge of the actual perspective of the speaker. This frame of reference precisely corresponds to the notion of egocentric perspective described in Chapter 5. A sentence stating that "the ball is to the left of the armchair" illustrates the deictic perspective on the scene that contains these two objects (Figure 10.4).

FIGURE 10.4 "The ball is to the left/to the right of the armchair"

Elements of the same repertoire of spatial terms can also be used when the system of coordinates is no longer centered on an observer/speaker, but on one of the objects of the scene. This is the case of the *intrinsic* frame of reference, in which coordinates are defined by taking into account the characteristics of the site-object (the armchair) to which the target-object (the ball) is related. This option is only possible for objects that have a natural "orientation," with an intrinsic front, back, left, and right side (buildings, edifices, vehicles, contrary to objects like trees, balls, stones, asteroids, etc.). Note that the sentence "The ball is to the right of the armchair," as produced within an intrinsic frame of reference, is still true even in the absence of an observer.

Lastly, while in the previous cases the object to which the statement refers is part of the scene (the observer in the first case, an oriented object in the second), the *extrinsic* (or *absolute*) frame of reference rests on the existence of arbitrarily fixed axes, generally cardinal directions. The reference is thus external to the scene and may be applied in the same conditions to any other scene. The position and orientation of the observer, or those of oriented objects present in the scene, are irrelevant when a speaker states, for instance, that "the ball is south-east of the armchair."

This classification of frames of reference is widely shared nowadays (see Levelt, 1996; Tversky, Taylor, & Mainwaring, 1997). There are, however, some variants. In particular, Stephen Levinson (1996) proposed reconsidering the classic notion of deictic frame and to substitute for it the notion of a *relative* frame of reference, which may involve any observer, including an observer distinct from the observer/speaker. The classification of the three classic frames of reference must then be redefined, with the relative frame implying a ternary relation, which includes the observer of the scene (Figure 10.5). While the man is "in front of the house" in an intrinsic frame of reference, it is justified within a relative frame to claim that the man is "to the left of the house," by taking into account the point of view of the observer, considering this observer's current position and orientation.

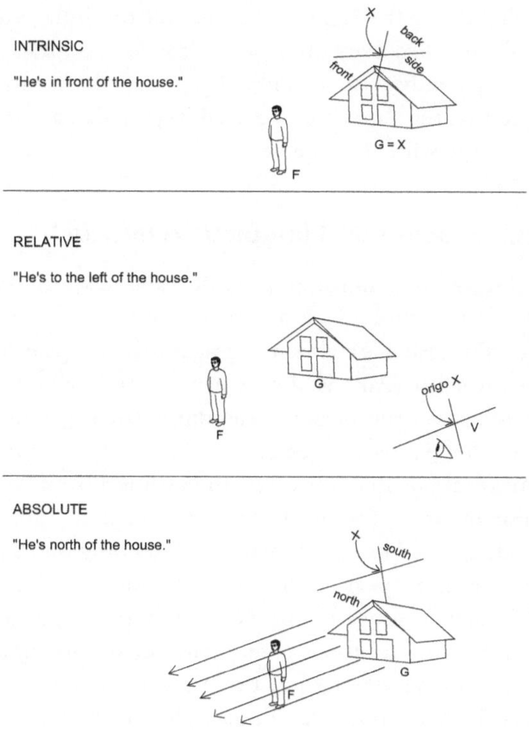

FIGURE 10.5 Three frames of reference. Reprinted from the book by S. C. Levinson, *Space in Language and Cognition: Explorations in Cognitive Diversity* (2003), by permission of the author and Cambridge University Press

Most speakers are able to use more than a single frame of reference and become adapted to other frames when a partner uses them to their intention. Each of these frames has advantages and disadvantages. The relative frame has the advantage of reflecting the point of view of the speaking person, but its inconvenience is to be based on a ternary system, obviously more complex. Binary relations used by intrinsic and absolute frames are simpler and do not require knowing the observer/speaker's point of view. But the intrinsic system requires that the partners share the same notion of what are the front and back of oriented objects. As to the absolute system, it supposes knowledge and mastery of cardinal directions by both partners (see Tversky et al., 1997).

Given the plurality of frames of reference and their respective advantages, one understands that all of them can be used in communication about space. Ideal communication might require that the choice of a perspective is consistent during the whole string of descriptive discourse. In fact, in many natural situations, speakers switch spontaneously from one frame of reference to another without the addressee seeming to experience special difficulty. A nice example, provided by Wolfgang Klein (1993), is the announcement delivered by a captain to passengers while the airplane is above Greece: "Greece is just beneath us. The big island beneath Greece, slightly to the right, is Crete, and the little white spot behind Crete is another island called Karpathos" (p. 195). As the message proceeds, the audience switches apparently without difficulty from a deictic to an intrinsic use of "beneath." Note that the captain could well have added a further reference to cardinal directions in the same message.

Comparative approaches and linguistic relativity

The comparative study of languages provides abundant illustration of their specificities in spatial terminology, sometimes in contrast with the uses of our Western languages. For instance, while the places that we speak about are generally considered as invariant (Paris is always named Paris, whoever is the person speaking of it and wherever this person is standing), the expression of toponymic knowledge may be affected by the speaker's subjectivity with a system of alternate names that makes it possible to express the position from which these places are considered. For instance, for the Inuit of the central Canadian Arctic, the same place can be designated by several names, depending on its position relative to the person who sees it or speaks of it (see Collignon, 2006). Another example is that of Wolof, which belongs to the Niger-Congo language family, and in which it is not possible to refer to an object without specifying its position relative to the speaker. Suffixes attached to the word designating an object specify whether the object that someone tells about is close to him/her or at a distance, whether the object is close to you but far away from him/her, or distant from both of you, but closer to you than to him/her (see Robert, 2006).

Another example brings us back to spatial prepositions. Vandeloise (1991) reported an interesting case about the reference to objects that have no intrinsic orientation. For an English or French speaker, a tree has no orientation that would allow identifying its intrinsic "front." The tree is in a way conceived as being posited "in mirror" to the observer (Figure 10.6, Scene 1). In the language spoken by the Hausa people, in West Africa, the situation is interpreted in a radically

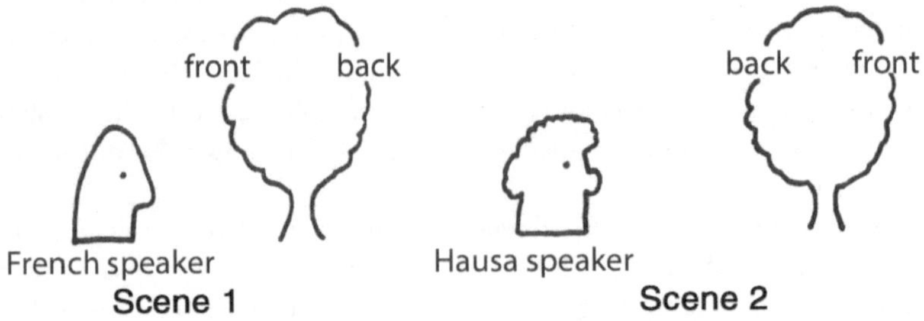

FIGURE 10.6 The front and the back of an object with no intrinsic orientation, for French and Hausa speakers. Adapted from the book by C. Vandeloise, *L'Espace en Français: Sémantique des Prépositions Spatiales* (1986), by permission of Editions du Seuil

different way. The tree is seen as having the same orientation as the observer ("in tandem"), in such a way that what is available to the observer's sight is the "back" of the tree (Figure 10.6, Scene 2) (see Hill, 1982).

The domain in which the most striking divergences are observed is that pertaining to the use of frames of references and the associated terminology. We noted that the deictic frame is generally considered as the dominant frame of reference, as being directly related to the egocentric perspective of people on their environment. This conception is mainly based on developmental arguments, in line with the Piagetian theory, according to which the capacity to adopt an absolute frame of reference is of late appearance, since it requires mastering the allocentric perspective and the linguistic codes that express it (see Piaget & Inhelder, 1967). However, studies conducted at the interface of linguistics and anthropology – in particular at the Max Planck Institute for Psycholinguistics in Nijmegen, the Netherlands – have pointed to language families for which absolute references (such as those attached to cardinal directions in Western cultures) are the first acquired and remain preferred by speakers of these far-off cultures.

Among the most striking examples is the Guugu Yimithirr, an Australian Aboriginal language from northern Queensland. In this language, all spatial descriptions refer to the equivalent of our cardinal directions. It is the case for the description of geographic spaces, but also of proximal spaces (objects displayed on a table, objects in a room). For instance, a description will state: "The lamp is north of the sofa and the table is west of it." The description will be the same whatever the position of the observer in the scene, whether the observer looks at it from the entrance door or from the window. Locative descriptions generated in this frame are invariant and are not affected by the rotation of the observer or of the object referred to linguistically. Therefore, to speak this language supposes having a permanent sense of orientation, knowing where cardinal directions are, and being able to compute the direction of any object on the basis of that knowledge. In sum, the speaker of Guugu Yimithirr is equipped with a "mental compass" which is continuously functional (see Haviland, 1998; Levinson, 2003).

In some other languages, the equivalent of cardinal directions to which a speaker refers is built around the topographical characteristics of the environment. In Bali, a geocentric orientation system is organized by reference to the sacred mountain, Gunung Agung. Orientation is geared either to the "upstream" direction (or "to the mountain") or the "downstream" direction (or "to the sea") (Figure 10.7) (see Wassmann & Dasen, 1998). Similar systems are found in other regions of the world. This is the case of Tzeltal, a Maya language spoken in the mountains of Chiapas, Mexico; Hindi, as spoken east of Uttar Pradesh, India; or the Nepali language of the mountainous region of Dolakha, Nepal, where directions are formulated by contrasting "uphill" ("to the mountain") and "downhill" ("to the valley") (see Dasen & Mishra, 2010; Hickmann & Robert, 2006; Levinson, 2003).

Here, we are facing the general issue of linguistic relativity and the question of the possibility for language to shape speakers' mental representations. First of all, in contrast with the Piagetian postulate, there is nothing universal in the traditionally

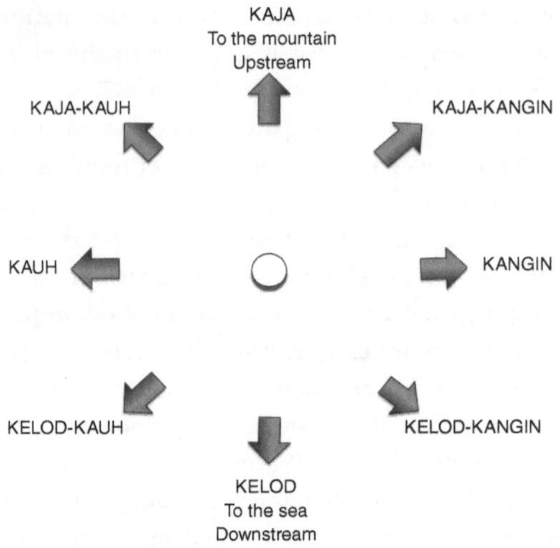

FIGURE 10.7 Balinese orientation system. Reprinted from the book by P. R. Dasen and R. C. Mishra, *Development of Geocentric Spatial Language and Cognition: An Eco-cultural Perspective* (2010), by permission of P. R. Dasen and Cambridge University Press

accepted chronology according to which an "absolute" spatial language and the associated representations would necessarily be conditioned by prior control of ego-centric language and representations. In children of many cultural areas, even if these are few in the world, the acquisition of an absolute spatial language (not restricted to the description of large-size geographic space) may well take place as early as that of a relative language.

Interestingly, people whose language leads them to adopt an absolute perspective in their spatial descriptions remain dependent on this perspective in many other situations involving space, even if they do not require any verbal production. Eric Pederson (1995) conducted insightful experiments with children from two Tamil sub-communities in the Madurai district of South India. One population was living in a social and cultural environment where descriptive language used the words "left"/"right"/"in front of"/"behind," while the spatial lexicon of the other population only referred to canonical directions "north"/"south"/"east"/"west." Relative coding for one group and absolute coding for the other were equally efficient, including for non-verbal spatial tasks (visual memory of geometric figures or spatial problem solving).

It should also be noted that early attachment to a particular frame of reference and the associated lexicon does not prevent people from demonstrating cognitive plasticity. A study by researchers from Nijmegen was conducted with children from the Netherlands, who spontaneously use relative (egocentric) spatial terminology, and Namibian children, whose dominant spatial language is geocentric. When children from either group are invited to follow instructions that use their non-habitual spatial language, they demonstrate their capacity to adapt themselves to this new language. But such adaptation has a cost, which is revealed by their lower performance in non-verbal spatial tasks (reconstruction of a configuration of several objects from memory), in comparison with children who receive instructions in their habitual spatial language (see Haun, Rapold, Janzen, & Levinson, 2011).

Having these data in mind, it becomes difficult to maintain a conception of spatial cognition that would universally build on the basis of the egocentric frame of reference. Children who grow up in an environment where they are exposed to absolute spatial language acquire this language as early as those who learn relative language. But a question ensues: if learning of languages associated with different frames of reference impacts verbal production and comprehension, might it also have deeper cognitive consequences?

We are ready to answer in the affirmative, while not ignoring the existence of prelinguistic spatial knowledge. There exist general features that are true of every environment where people grow up (gravity, object permanence, etc.). There are also cognitive schemas that preexist to language and are likely to be universal (the relation of containment, the relation of support, etc.) (see Bowerman, 1996; Levinson, 1996; Pederson, 1995). But this reality does not invalidate the empirical arguments in favor of the Whorfian theory that as early as language is settled, it constrains cognitive processes and contributes to formatting human cognition (Whorf, 1956). The arguments in favor of the role of language in the structuration of spatial cognition remain very powerful (see Majid, Bowerman, Kita, Haun, & Levinson, 2004).

Beyond the arguments provided by comparative psycholinguistics, interfacing between spatial language and the expression of spatial knowledge and capacities is attested by developmental studies. For instance, it is well established that children whose parents make use of large amounts of spatial vocabulary have linguistic productions that are richer in spatial terms than children of the same age less exposed to spatial terminology. Still more remarkably, these children are more likely to perform better at a later age in tasks requiring good control of visuo-spatial capacities, such as figure transformations or discovery of spatial analogies (see Pruden, Levine, & Huttenlocher, 2011). In complement to sensorimotor experience, spatial language does contribute to improving the quality of people's spatial thinking.

11

SPATIAL DESCRIPTIONS

It is classic to point to the contrast between the multidimensional structure of space and the unidimensional organization of language. Whereas spatial objects are distributed over at least two dimensions, it is inherent to language to convey information in successive pieces, according to sequences governed by a set of syntactic rules. By construction, the full set of statements that are combined to verbally account for a situation can only be delivered sequentially. This constraint obliges the speaker to decide the order in which the components of a linguistic message will be stated and thus communicated to an interpreting agent. This is true of every subject of discourse, but in the case of space, a set of particular decisions will have to be made with the objective of finding a sequence that gives the addressee the possibility of building a coherent and informationally valid representation.

To start with, from which point of a configuration, from which object in a scene should a description be initiated? Then, which further object should be posited by reference to the first one, and so on? For even a small set of objects that compose a scene, there is a huge number of theoretically possible sequentializations. This situation creates a challenge for the art of description, although it soon appears that a narrower set of options is eventually retained by speakers/writers. Which factors constrain describers to finally converge on a limited number of descriptive sequences?

To describe a scene is not a self-evident exercise. The activity is likened to the resolution of the so-called "ill-defined" problems, that is, problems which do not allow for a single solution, but several possible ones. These solutions, in the case of spatial descriptions, can be hierarchized and classified as "more or less" adequate in terms of clarity and communicative value. Will we identify some descriptive sequences as more sound than others to communicate information and optimize the chances of an addressee acquiring new knowledge in good cognitive conditions?

Literary challenges

Let's begin to explore the issue of descriptive linearization by looking at decisions made under the pen of writers. In Chapter 3, we have mentioned some encounters between literary art and the expression of spatiality. Description is integral to literature. By referring to this domain, we are not outside the field of cognition. Writing is an activity aimed at a reader. The writer is guided by the intent of having a reader create representations, which are built and enriched as the narration unfolds, and possibly remain in the reader's memory long after their construction.

Description is hardly seen as a literary genre on its own, but it is a building block of various forms of literary creation. It is typically combined with narration, as illustrated by the great tradition of novels in the nineteenth century, but also travel narratives, equally popular around the same period.

A revealing example of challenges imposed by descriptive writing has been provided by Gustave Flaubert. As a novelist, Flaubert was known to be extremely rigorous as regards the quality of his writing (see Winock, 2013). His manuscripts, kept by the Bibliothèque Nationale de France, attest of his strenuous efforts, in continuous and quasi-obsessional search of the right and most finely cut expression of each sentence. The successive steps of the writing of his story *Hérodias*, the last of *Three Tales* (1877), have been preserved in the form of fourteen successive versions of the manuscript, produced during a writing period of four months.

The first paragraph of the text is devoted to the description of the citadel in which the story is about to take place:

> In the eastern side of the Dead Sea rose the citadel of Machærus. It was built upon a conical peak of basalt, and was surrounded by four deep valleys, one on each side, another in front, and the fourth in the rear. At the base of the citadel, crowding against one another, a group of houses stood within the circle of a wall, whose outlines undulated with the unevenness of the soil. A zigzag road, cutting through the rocks, joined the city to the fortress, the walls of which were about one hundred and twenty cubits high, having numerous angles and ornamental towers that stood out like jewels in this crown of stone overhanging an abyss.

While the reader of this final, eventually published version has a sense of fluent writing which unfolds naturally, the genesis of these first lines necessitated a huge amount of rewriting work on the part of Flaubert. The incipit of *Hérodias* has been thoroughly analyzed by specialists of the study of manuscripts (Grésillon, Lebrave, & Fuchs, 1991). The analysis reveals, first of all, the efforts of the writer to work on the basis of a visual representation of the citadel, as precise as possible, with the help of drawn sketches, in order to translate this representation into text and thus proceed "from mental to verbal." Moreover, the sheets showing the fourteen successive versions of the paragraph (first, as notes in telegraphic style, then repeated attempts of textualization) reveal endless successions of corrections,

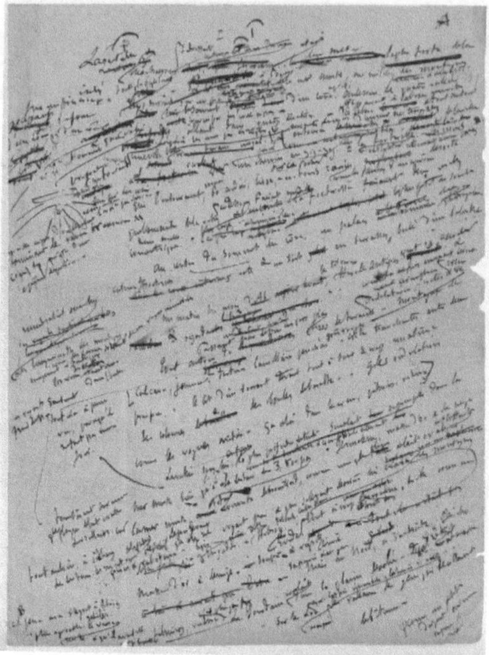

FIGURE 11.1 First page of the ninth version of Flaubert's manuscript. Reproduced with the permission of the Bibliothèque Nationale de France

adjustments, reformulations, paraphrases, additions, and deletions, which attest the extent to which, even for a highly recognized writer with exceptional literary talent, description may be a costly and painful exercise (Figure 11.1).

Linearization

Linearizing a verbal description – text or discourse – does not entail any difficulty when the reality to describe has a structure that naturally guides the describer. For instance, the description of an itinerary will predictably begin by some reference to the starting point, then will review the successive steps until the destination point. The situation sharply differs in the case of the description of a static configuration containing several objects. If the scene is not organized according to a temporally constraining structure, a wide variety of sequences may be used by describers. However, it is interesting to look for dominant discursive structures likely to reflect consistent modes of propositionalization of a scene. Charlotte Linde and William Labov (1975) asked New York City residents to orally describe their flats. It appeared that the spatial representation of this kind of familiar environment was spontaneously transformed by the majority of informants into a temporal sequence (that the authors termed "pseudo-narrative"). The process is obviously governed by implicit rules. Not only is it infrequent to find descriptions made from a survey point of view, as if the person described the layout of the apartment, but remarkably, describers start their "walking tour" from the entrance door. The transition from a cognitive representation to a verbal sequence draws on a temporal structure which reflects a virtual displacement through the described environment.

The situation is quite different when, rather than accompanying a simulated locomotor displacement, the description is about a scene whose elements are all simultaneously available to the sight of the observer/describer. Veronika Ehrich and Charlotte Koster (1983) asked people to look inside a doll's house through its main door and give an oral description of the pieces of furniture contained in the house. The results showed that people generate descriptions that take the form of a "gaze tour" by moving along the walls, with a strong tendency to start descriptions on the left and proceed rightwards. The regularity of some descriptive sequences is also found when students are required to describe a familiar room from memory (see Shanon, 1984).

An interesting experimental situation, although a very simple one, was proposed by Pim Levelt (1982, 1989), in the form of geometric figures (colored circles) connected by horizontal and vertical lines (Figure 11.2). The describers are invited to provide a description to a (virtual) person who is not exposed to these figures, but will have to reproduce the pattern with color pencils. No constraint is imposed on the describer, except as regards the starting point of the description (the first circle

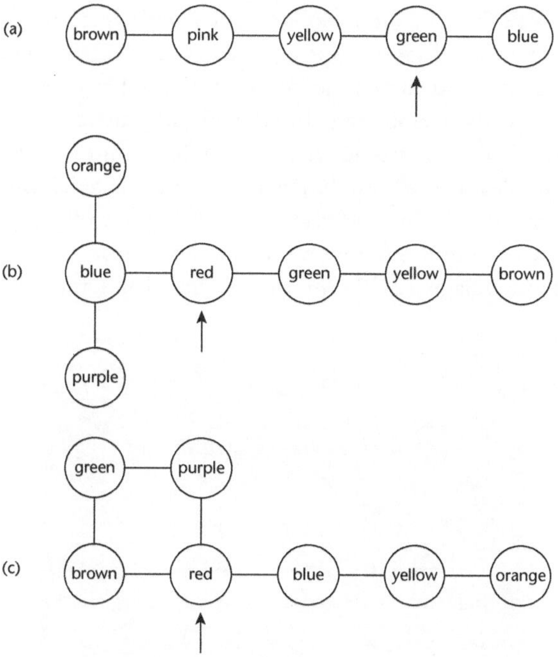

FIGURE 11.2 Figures used in W. J. M. Levelt (1982) and F. Robin and M. Denis (1991) experiments. Reprinted from the chapter by M. Denis in *Models of Visuospatial Cognition* (1996), by permission of Oxford University Press

to be named is the one marked with an arrow). Levelt showed that describers adopt linearization strategies which are far from random. In general, the dominant strategy consists in first describing the branch which minimizes the number of items that describers have to store simultaneously in working memory (or the duration of such storage). For instance, in the description of the second of the three figures, two thirds of the participants mention first the linear branch (here, to the right), then the embedded branch (here, to the left). Further studies showed that when people have to describe mental images constructed from such figures, they adopt the same linearization strategies as those used for the description of the same configurations available to their sight (see Robin & Denis, 1991).

The works inspired by the theory of mental models (Johnson-Laird, 1983) also contributed to revealing the linearization processes brought into play in descriptive discourse. They evidenced the impact of the organization of such discourse on its comprehension and memorization.

In that respect, in order to help someone construct the representation of a simple configuration of a few objects, there is no need for a description to exhaustively assert every spatial relation between all pairs of objects. Consider a collection of five objects on a table (Figure 11.3). For a person invited to imagine sitting in front of the table, only four statements (instead of ten) will allow him/her to confidently reconstruct the configuration (under the assumption of equal distances between adjacent objects): "(1) The melon is to the left of the lemon. (2) The pear is to the right of the lemon. (3) The banana is in front of the melon. (4) The apple is in front of the lemon." Such a discursive option reflects descriptive parsimony, which of course supposes that the describing person has identified the necessary and sufficient subset of expressible relations that define without ambiguity the content and structure of the configuration.

A principle consists of privileging linearization that ensures referential continuity of the description. Ideally, each time a new item is introduced, it should be referred to a previously introduced one and, whenever possible, to the immediately prior mentioned one. Consider the following spatial description: "(1) At the center of the island, there is a village. (2) To the east of the village, there is an aerodrome.

FIGURE 11.3 A configuration of five objects

(3) To the north of the aerodrome, there is a plantation. (4) To the east of the plantation, there is a hill." This description conforms to the principle of referential continuity. Suppose that in a variant of this text, Sentences 3 and 4 are inverted. This variant essentially provides the same pieces of information as the previous version, but when readers who have just processed Sentence 2 are presented with Sentence 4, they have no possibility of integrating the new piece of information conveyed by Sentence 4 in the representation they are currently constructing. It is only when Sentence 3 is processed that such integration can be achieved. Holding a piece of information as an isolated item while expecting further information that makes it possible to integrate it in the mental representation under construction has a cognitive cost, which is likely to interfere with current processing and further operations (see Denis, 1996; Denis & Denhière, 1990; Ehrlich & Johnson-Laird, 1982; Foos, 1980; Wagener & Wender, 1985).

Another well-identified principle consists of privileging descriptions that avoid the reader or listener building simultaneously alternative versions of the same content. If we take again the description of five fruits, let's introduce a very small change, by replacing the second sentence by this new sentence (also true): "The pear is to the right of the melon." The description now causes ambiguity for the reader or listener since it is not only compatible with the previously considered arrangement, but also with another, distinct one (Figure 11.4). This descriptive option creates an indetermination that can only be alleviated by introducing at least one additional statement. In the absence of it, building two distinct mental models, both compatible with the descriptive formulation, creates an adverse situation in terms of cognitive load (see Byrne & Johnson-Laird, 1989; Mani & Johnson-Laird, 1982).

Selecting a perspective

Several perspectives may be taken on an environment, in particular when a describer intends to help a reader or listener to build a representation – a "mental model" – of the environment in question. The most frequently highlighted contrast is between

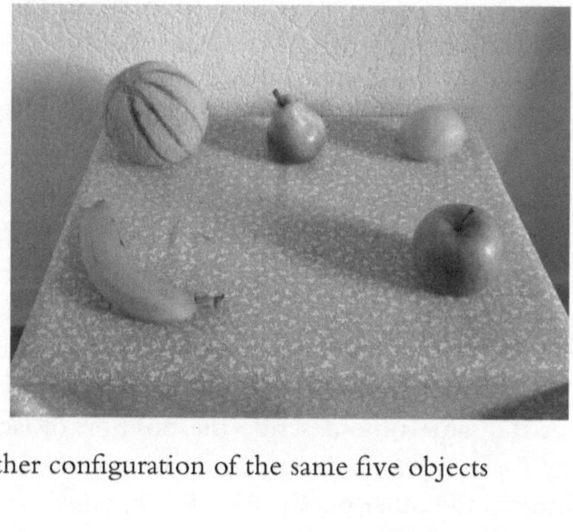

FIGURE 11.4 Another configuration of the same five objects

"route" and "survey" descriptions. Holly Taylor and Barbara Tversky (1992) have illustrated these two perspectives with descriptions of urban environments. A tourist guide favoring route descriptions would invite the reader to move mentally as follows:

> As you sail up the Seine from the Place de la Concorde, you first come to the Musée d'Orsay on your right, and then the Louvre on your left. Straight ahead of you, you can see the Ile de la Cité, and the spires of Notre-Dame . . .

On the other hand, a description adopting a bird's-eye view would say:

> The Washington Mall is bounded by the Capitol at the east and the Lincoln Memorial at the west. Museums line the southern and northern borders of the middle of the Mall. Along the southern part of the Mall, the most eastern museum is the Air and Space Museum. Just west of it, across the street, is the Hirshhorn Museum . . .

A route description leads a reader or listener to adopt the point of view of someone moving through an environment. The navigating person is encouraged to take a frontal, egocentric view, which will inevitably change as reorientations are executed along the path. Route descriptions preferentially use egocentric language ("on your right," "on your left," etc.). Contrastingly, survey descriptions impose on the reader or listener a view from above which makes it possible to apprehend the described scene as a whole. The allocentric perspective induced by this type of description is likely to be maintained all along the description. Spatial terminology largely resorts to cardinal directions ("to the east," "to the west," etc.).

Given the sharp differences between the two perspectives, the first question to raise is whether the discourse that they elicit generates different mental representations in their users' minds. The first approaches to this issue brought answers evidencing such differences, during the course of processing as well as in memory representations (Perrig & Kintsch, 1985). Taylor and Tversky (1992), on their part, committed themselves to stress the features common to representations built from these two sorts of descriptions. To this aim, they invited people who had read either a route or survey description to verify statements about spatial relations that had not been asserted in the text, but could be inferred from the information conveyed by the text. The results showed that judgments on these inferences were as fast and correct, whichever the perspective imposed by the description or the perspective taken by the statement of the to-be-judged spatial relation. These results indicate that mental models constructed from these two sorts of descriptions can be abstract and schematic enough to encode spatial relations stated from both perspectives.

The duality between route and survey perspectives has been a continuous source of inspiration since Taylor and Tversky's (1992) original studies. The same authors also showed that besides "pure" descriptions (route or survey), there are many examples of "mixed" descriptions, such as those which lead a person virtually along a route and at the same time describe the position of landmarks by using cardinal terminology. In most cases, describers do not even warn their addressees that they are switching to the other perspective. For instance:

> The town is south of the White Mountains which run east–west. From the mountains, Mountain Road runs directly south into town. At the first road in town, a right turn takes you past a store to the right, then a left turn takes you past a school to your right again . . .

Route describers can also switch to a survey description, then return to the descriptive mode by which they first began. Another noteworthy aspect of perspective choice is that they are to a large extent determined by characteristics of the environment to which the description applies. For instance, environments traversed by a single path strongly encourage route descriptions, whereas open environments with multiple paths and landmarks more frequently elicit a survey perspective (Taylor & Tversky, 1996).

Works on the compared qualities of mental models built from the two sorts of descriptions have flourished. First, the notion of quasi-indifferentiation between representations issued from route and survey descriptions has been questioned. In particular, it was shown that readers of a survey description are more prone than those of a route description to draw sketch maps when they have to recall the information they have memorized, thus attesting that they hold an integrated spatial representation of the information conveyed to them (see Schneider & Taylor, 1999).

Furthermore, although empirical evidence is available that information about distances can be coded by listeners of route and survey descriptions, the latter nevertheless may be of better quality, if only because they generate more fine-grained representations than route descriptions (see Noordzij & Postma, 2005). Other studies involving mental comparison of distances confirm that metric information is less accurately encoded in representations issued from a route description (see Péruch, Chabanne, Nesa, Thinus-Blanc, & Denis, 2006). Lastly, switching to another perspective is easier for people whose initial learning was about a survey rather than a route description (see Brunyé, Rapp, & Taylor, 2008).

As to the role of working memory, it appears that distinct components thereof are called upon by the two perspectives. Admittedly, visuo-spatial memory is particularly solicited in the processing of texts with spatial content. For instance, a secondary spatial task interferes with memorization of texts describing spatial environments (such as a zoo or a tourist center) (see Pazzaglia & Cornoldi, 1999; Pazzaglia, De Beni, & Meneghetti, 2007). But it is primarily the *spatial* component of visuo-spatial memory that is implied in the processing of descriptions which adopt (and invite the readers or listeners to adopt) a survey perspective, compared with route descriptions. Lastly, in line with the findings reported in Chapter 4 on the role of working memory in navigational performance (Garden et al., 2002; Meilinger et al., 2008), memory of route descriptions is particularly sensitive to interference resulting from a concurrent visuo-spatial task (see Brunyé & Taylor, 2008; Deyzac, Logie, & Denis, 2006; Pazzaglia, Meneghetti, De Beni, & Gyselinck, 2010).

Taking individual differences into account reveals that some people have a marked inclination for route perspective while others prefer survey perspective, according to their responses to self-evaluation questionnaires. The former claim that they spontaneously create representations of their environment by revisiting the paths that they have followed, by using an egocentric frame of reference. The latter report that they represent the environment in the form of a mental map and more willingly refer to cardinal directions.

It turns out that people who describe themselves as moderately inclined to adopt a survey view have poorer navigational performance when they have to learn an environment from a survey perspective (from learning a map). Contrastingly, those who favor survey strategy commit fewer navigational errors, whatever the learning perspective imposed on them, this showing their moderate sensitivity to the effects of perspective (see Pazzaglia & Taylor, 2007). The same participants, who are at ease with the survey perspective, perform better than the others in the verification of sentences containing spatial inferences, in graphic recall of the described environment, in directional pointing to remote places, as well as in mental rotation. They also demonstrate their capacity to build a survey representation from a route description, of which they take advantage in the execution of further tasks in the same environment (see Meneghetti, Pazzaglia, & De Beni, 2011; Pazzaglia & Meneghetti, 2012).

"Moving" through a mental spatial model

The reader of a spatial description is thought to build, as a complement to the propositional representation of the text, a representation of the *situation* to which the text refers. This representation has been characterized as a "mental model" in the theory of Johnson-Laird (1983) and a "situation model" by van Dijk and Kintsch (1983). However, the two conceptions have in common a consideration that the model is not organized according to a sequential structure as is the case of language, but that its structure is analogous to that of representations derived from perception.

The mental model built by a reader may well include pieces of information that have not been made explicit in the text, but are inferred from it. These elements result from the reader's interpretation of the situation that the writer had in mind while he/she was describing the scene and that he/she intended to share with the reader. Each term has to be interpreted in the context of the described situation. In particular, it is frequent that a text describes scenes without specifying any detail about the spatial relationships among the objects. A reader may well create a mental model endowed with a dimensionality such that these relationships are expressed in it. An illustrative example of this phenomenon has been reported by Daniel Morrow and Herbert Clark (1988), in an experiment where participants had to read statements, for instance the description of a person sitting at a sidewalk cafe looking across a plaza at a cathedral with a statue in front of it. The text followed by introducing a character, namely a nun, who was said to be "approaching the statue" in one version and "approaching the cathedral" in another. When the participants were questioned about the described situation, those who read the first version estimated that the distance between the nun and the statue (7 meters) was shorter than the distance between the nun and the cathedral in the second version (12 meters). The notion of "nearness," which allows us to circumscribe the region where interaction between an agent and an object takes place, appears to be modulated by the size of the entities involved in the situation. As a result, for two descriptive sentences of the same structure and complexity in the text base, the respective mental models effectively differ from each other.

An abundant experimental literature has been devoted to studying how readers follow a narrative episode that unfolds in the representation they have previously constructed of an environment. In the paradigm developed by Daniel Morrow and his colleagues (Morrow, Bower, & Greenspan, 1989; Morrow, Greenspan, & Bower, 1987), readers are first invited to study a diagram of a building composed of several rooms, each containing several objects, in which the story will take place (Figure 11.5). The experiment proper begins after checking that the participants have accurately memorized the rooms and the corresponding objects. Participants are then presented with a text describing a person moving through the building to achieve a goal. For instance, in the case of a research institute, the story is about the director making a tour to check whether everything is in order before a committee's visit. The text includes critical motion sentences specifying the move from a given room to an adjacent one (for instance, "Wilbur walked from the conference room into the laboratory"). After each critical sentence, the narration is briefly interrupted by the presentation of two object names. The participants have to decide whether the two objects come from the same or different rooms. Reading is then resumed until the next critical motion sentence and a further question about two objects.

The idea tested in this experiment is that during the reading of a narrative taking place in an environment, all the regions in the mental model are not activated to the same degree. In particular, the point of space where the character is currently staying is thought to be at the highest level of activation. As a result, the objects

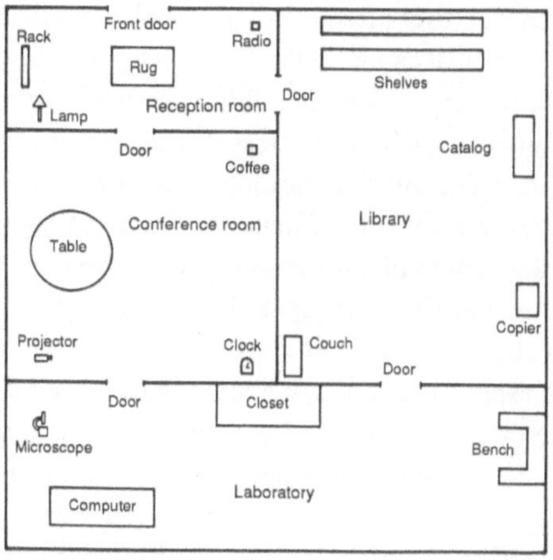

FIGURE 11.5 Diagram of a building. Reprinted from the article by D. G. Morrow, S. L. Greenspan, and G. H. Bower (*Journal of Memory and Language*, 1987), by permission of Elsevier

located near the character are cognitively more accessible than are more remote objects. This assumption is confirmed by the fact that decision times about an object pair are shortest when these objects are in the room the character is walking into (in this example, the microscope and the computer). Longer times are necessary for object pairs in a room that the character has just left (the table and the projector). Times are still longer when the objects are in another room (the couch and the copier) or another building. Metaphorically, everything happens as if the moves of the character were followed by a spotlight circumscribing an area of maximal activation, and thus higher accessibility. Activation of this area decreases when the character moves away to reach another region of the modeled space. The existence of a *gradient of accessibility* supports the idea of a moving focus induced by the narrative. The distance effect shows that activation in the model is not regulated by an all-or-nothing process, but by a gradual mechanism controlled by the changing positions of the character.

These findings need to be modulated by two considerations. First, the question is to establish at which degree of detail a spatial model must be constructed to be useful to readers, given the task expected from them. In natural circumstances of reading, the building of a model endowed with quasi-metric characteristics is neither the most spontaneous, nor the most relevant strategy. But when a specific objective is assigned to readers (here, answering questions about objects' localizations), the task will require the implementation of a detailed mental representation. This would probably not be the case if the context was that of reading "just for pleasure." There are degrees of precision in a spatial model, as a function of the requirements of the task associated with reading (see Gray Wilson, Rinck, McNamara, Bower, & Morrow, 1993).

Furthermore, the notion of a spatial gradient of accessibility leads us to wonder about a metric that would be inherent to spatial mental models. In fact, Euclidean distances (as measured on the maps studied by the participants) do not affect response times. The spatial distance effect is in fact essentially determined by the *number of rooms* separating the point on which readers' attention is focusing and the objects to be considered. When the sizes of the rooms are redesigned in order to modify their metric, accessibility is not affected (see Rinck, Hähnel, Bower, & Glowalla, 1997). However, an effect of Euclidean distance is evidenced when readers, instead of imagining the moves of a protagonist from the narrative, receive imagery instructions that invite them to represent their *own* moves in the environment. This finding suggests that mental imagery incites readers to engage in a finer and more precise mental representation that incorporates Euclidean distances, rather than a coarse representation in terms of the number of traversed rooms. Clearly, in some contexts, mental simulation does not necessitate generating very sharp representations. In any case, mental imagery provides a means of creating specific and detailed representations, which are responsible for both effects (number of traversed rooms and Euclidean distances) (Rinck & Denis, 2004).

Quite interesting are the efforts to elucidate whether a reader is able to represent distinct perspectives held by several characters within a described environment, and

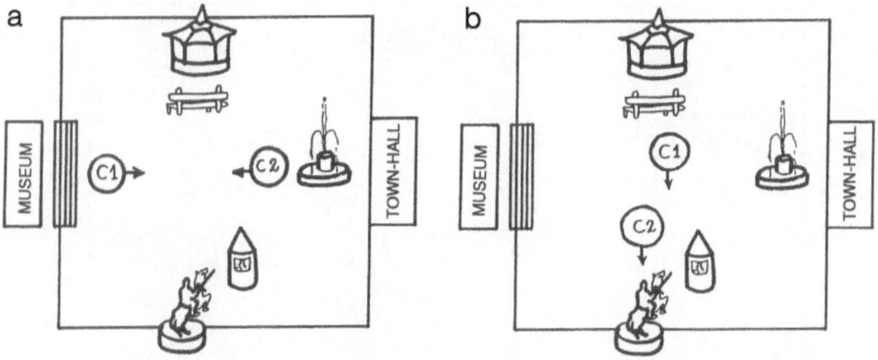

FIGURE 11.6 Scenes used in experiments by M. de Vega (1994). C1 and C2 represent the priest and the postman, respectively. Reprinted from the chapter by M. Denis in *Models of Visuospatial Cognition* (1996), by permission of Oxford University Press

to switch from one perspective to another while reading. To this question, Manuel de Vega, from the University of La Laguna, provided a highly informative answer. His objective was to question the idea that when several characters are involved in a narrative, readers are not inclined to adopt in turn the point of view of each character, but they rather stick to a neutral perspective that encompasses the whole scene (Franklin, Tversky, & Coon, 1992). In the situation designed by de Vega, the participants memorize the description of a scene that includes several objects and in which two characters (here, a priest and a postman) are involved (Figure 11.6). When participants are invited to verify sentences describing the relation between a character and an object in the scene, they demonstrate that they are able to adopt different points of view, but these perspective changes have a cognitive cost. For instance, after reading introductory sentences that place the focus on the priest, they are faster at verifying the sentence "The priest has the kiosk on his left" than the sentence "The postman has the kiosk on his right" (Figure 11.6 a). Furthermore, when the description has introduced the two characters with the same point of view on the scene (Figure 11.6 b), verification responses are faster than if the description mentioned that the two characters had different points of view (see de Vega, 1994).

Equivalences?

There is no doubt that language is a valuable source of spatial knowledge. However, if spatial information can be acquired from verbal descriptions, is it comparable to the information acquired from direct experience of an environment or its map representation?

Although large amounts of information can be conveyed by language, neither describers nor receivers ignore the limits inherent to spatial language. Even by pushing descriptive efforts to extremes (as in Lovecraft's description of old Quebec

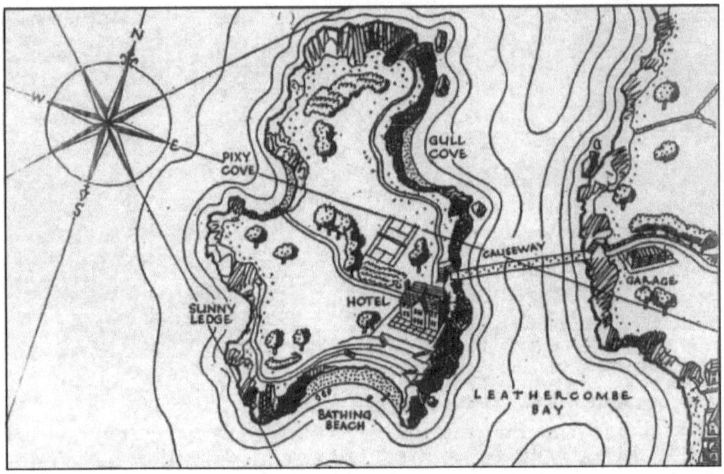

FIGURE 11.7 Diagram provided at the beginning of Agatha Christie's novel, *Evil Under the Sun*. Reprinted by permission of HarperCollins Publishers Ltd © Agatha Christie (1941)

City, quoted in Chapter 3), a number of aspects cannot be captured, and thus rendered by language. Even if the effort was pushed to its limits by a highly committed expert describer, the processing of the resulting output would soon turn to a cumbersome cognitive experience for the reader. This is the reason why some writers who want to save time and effort for their readers provide them with a diagram of the environment in which the story is about to take place, sometimes on the very first page of the book. By doing this, they spare the burden of endless descriptions and allow readers to readily grasp the spatial frame whose details will be significant for good comprehension of the story. Examples can be found in Agatha Christie's novels, which provide readers with the opportunity to create a mental model in which the coming crime story will take place (Figure 11.7). In sum, much ahead of her time, the British novelist provided an equivalent of what Morrow's experimental situation delivered to the participants.

If we now consider situations devoid of any possibility for the describer and the reader/listener to rely on figurative media, we must envisage the opportunity that mental imagery offers to encode visuo-spatial information from an exclusively verbal input. If the description is detailed enough and conveys geometric information of moderate complexity, it should be an option to have the reader/listener build the mental figuration of an object or a configuration of objects, which contains spatial information structured as the object or the configuration themselves.

This has been precisely the objective of a research program conducted in our CNRS lab, in Orsay, in order to document how the human mind represents metric information contained in spatial configurations that have been learned from a verbal description, without any perceptual contact with the configurations.

The experiments are based on the paradigm originally developed by Stephen Kosslyn to demonstrate the structural isomorphism of visual images with respect to previously learned configurations, and thus the analog character of this particular type of representation. When a person has memorized the map of an environment and is keeping in memory a reliable mental representation of the configuration, mental exploration of the distances between landmarks reveals an interesting regularity. Mental scanning times linearly increase as a function of inter-landmark distances in the configuration. This correlation, which by the way is not in line with common intuition, is interpreted as reflecting that metric properties of the configuration are effectively preserved within the representation (see Denis & Carfantan, 1985; Denis & Kosslyn, 1999; Kosslyn, Ball, & Reiser, 1978).

In our adaptation of the paradigm, the learning of the configuration was only based on a verbal description delivered a sufficient number of times to ensure that a full and consolidated representation of the configuration was available to the participants. The description was about an island circular in shape, with six geographical features situated at its periphery. The locations of the features were indicated verbally in the conventional hour-dial terms. For instance: "At 11 o'clock, there is a harbor. At 1, there is a lighthouse. At 2, there is a creek, etc." The participants were then asked to mentally scan between all pairs of landmarks. The results showed that scanning times were proportional to actual distances between landmarks, just as they are when participants perform mental scanning after learning of the map itself. These results suggest that the spatial model constructed by the participants contains veridical metric representation of the described figural space (see Denis & Cocude, 1989).

The remarkable aspect of these findings is to attest that mental representations available after learning contain *more* information than that explicitly delivered by the description. Not only does the representation contain information about the position of geographic landmarks as stated in the description, but it also contains information on the relative *distances* between them, although the descriptive text did not mention anything on that and the participants had no access to such information. This state of affairs derives from a critical property of spatial models, that is, when a description specifies the absolute positions of the entities that compose the configuration, their relative positions are, by construction at the very same time, made evident. On the whole, distances are thought to be effectively represented in the form of an internal metric, and this metric supports valid computations, as in mental comparison of distances (see Denis, 2008; Denis & Zimmer, 1992).

Building a spatial model is not an all-or-nothing process. It is gradual, and the metric is fully expressed only after a sufficient amount of learning. Elaboration of the representation goes through successive steps, with progressive decrease of the initial "fuzziness" of the representation and correlative increase of its "sharpness." Besides, in line with what we mentioned above, the order in which information is delivered by the descriptive message impacts the construction of the representation. Referential continuity is a facilitating factor of this construction (see Denis &

Cocude, 1992; Denis, Gonçalves, & Memmi, 1995). Provided a sufficient amount of learning is secured, the mental model built from the spatial description of a limited set of elements is likely to incorporate reliable metric information.

Interestingly, neuropsychology and neuroimaging reveal similarities of brain functioning in situations where information is either presented perceptually or verbally. For instance, behavioral indicators of spatial neglect observed in parietal patients for perceived or remembered visual scenes are also found in experimental settings where the arrangement of objects in a scene has been conveyed in a purely verbal form (see Denis, Beschin, Logie, & Della Sala, 2002). PET also reveals activation of the same parieto-frontal network when people mentally explore spatial representations built after map learning or listening to a description. However, beyond such similarity, the angular gyrus and language zones are activated when people evoke mentally spatial information derived from the description. The human brain seems to keep track of the verbal origin of memorized spatial information (see Mellet, Bricogne, Crivello, Mazoyer, Denis, & Tzourio-Mazoyer, 2002; Mellet, Tzourio-Mazoyer, Bricogne, Mazoyer, Kosslyn, & Denis, 2000).

The mental scanning paradigm has mainly been used for texts describing proximal environments, those pertaining to figural space. However, the issue of similarity between representations based on direct experience and those mediated by language must also be considered in the case of spatial environments of larger extent. Researchers from Santa Barbara and Carnegie-Mellon have investigated this question. When people are in an open space, what happens if they are instructed to move to a target whose location is described verbally to them, for instance "2 o'clock, 5 meters"? Measures of azimuth and distance covered during walking by blindfolded participants reveals that performance is similar to a condition where moves are guided by a sound signal emitted from the target location (before initiating the move). Other similarities also appear when the displacement towards the target is from a starting point different from the location where sound or verbal information has been delivered and thus implies updating of the target's location. The quasi-equivalence of performance suggests that when an internal representation of the target's place has been created, subsequent performance is independent of the type of information used to build that representation. Similar patterns are also found when visual modality is involved instead of sound signals (see Klatzky, Lippa, Loomis, & Golledge, 2003; Loomis et al., 2002).

However, if language helps in generating spatial images likely to be updated if necessary, their functional equivalence with images built from other sources does not imply that these representations are as easy to build from every modality and are ultimately identical in nature. In some particular conditions, the postulate of equivalence may not be verified. Let us consider the situation in which people learn the positions of four objects in a room either from viewing the scene or from a verbal description ("The cat is at 3 o'clock, at 3 meters"). When people are later asked about spatial relations that have not been made explicit during learning, their allocentric judgments of direction and distance (for instance, in which direction is the cat from the place where the baby is?) are correct, but directional responses are slower and show larger variability than when learning has involved visual modality (see Avraamides et al., 2004). Thus, the assumption of functional equivalence has limits, since it cannot be attained in tasks involving complex environments and requiring particular cognitive computations.

Other explorations of the same issue deal with language as a vehicle of expression of spatial knowledge. Manuel de Vega and María José Rodrigo (2001) dedicated themselves to this question. They started from the classic data showing that updating spatial relations between objects that surround an observer (by digital pointing, blindfolded) is easier when the person physically rotates him/herself than when invited to simply imagine oneself performing such rotation (see Rieser, 1989). The classic interpretation of this finding invokes the fact that proprioceptive feedback elicited by the observer's physical movement ensures automatic updating of the new positions of objects around him/her. In de Vega and Rodrigo's experiment, where participants were sitting on a rotating chair and were surrounded by four objects, the same relation was found (pointing responses were given by pressing one of four directional arrow keys). It turns out that pointing is faster and more accurate when participants have physically rotated than when rotation has been imagined. But the newest information here is that this difference is not found when participants give directional responses verbally. Imagined rotation does not seem to engender particular difficulty as compared to physical rotation. This difference forces us to make a distinction between two types of spatial representations: those which are anchored in an individual's current sensorimotor environment and those which are detached from sensorimotor information. The latter would be inscribed in a purely "representational" frame, which would not take profit of automatic updating made possible by physical moves in the environment.

Marios Avraamides, from the University of Cyprus, extended the investigation of updating of spatial relations when reorientation movements are imagined (rather than performed physically). In a learning situation where the representation of an environment – more complex than in de Vega and Rodrigo's (2001) experiment – is fully constructed from a descriptive text, responses following physical rotation appear to be both faster and more accurate than those following imagined rotation. In the latter case, once the new (imagined) perspective is adopted, the absence of proprioceptive information requires more cognitive effort (Avraamides, 2003).

With language, people can thus create the representation of a remote environment, detached from immediate experience, involving a coordinate system that is not necessarily aligned with their current (egocentric) frame of reference. One can then understand that updating of people's orientation towards distant environments, even familiar ones, is not achieved as easily as towards a close environment, even in the absence of visual information (see Wang, 2004). When judgments are made on spatial relations between objects in a distant environment that has been described verbally, people who have to accomplish mental perspective changes have difficulty in detaching themselves from the initial orientation they have taken on the described environment. Although people have the capacity to build veridical spatial models from verbal descriptions and perform computations and updating (within a moderate level of complexity of described environments and tasks), it seems that in-depth differences remain between representations of environments built from language and representations of perceptually experienced environments (see Avraamides, Galati, Pazzaglia, Meneghetti, & Denis, 2013). Interestingly, people demonstrate their ability to combine visual and verbal information about the same environment when their attention is drawn to the cognitive advantage of creating an integrated model of this environment (see Pantalides, Kelly, & Avraamides, 2016).

12

ROUTES AND ROUTE DIRECTIONS

Route directions are a specific type of spatial discourse, whose aim is to help a person move through an unfamiliar environment and reach a target. As with any form of discourse, it conveys information, but a type of information that is intended to elicit behavior in the form of a navigational episode (walking or aboard a vehicle). When travel is completed, the person has reached a destination and, in some cases, retains in memory the information that was used to that effect.

Discourse intended to assist navigation belongs to the more general class of procedural discourse, in other words, discourse delivering instructions that would elicit an individual's action (or sequence of actions). Other variants of procedural discourse are illustrated by assembly instructions, operating instructions, cooking recipes, etc. They share the characteristic of describing the steps to follow in order to achieve a goal. The actions specified in procedural prescriptions have a critical chronological dimension, particularly evident in the case of route directions, since these instructions stick to the very sequence of actions that they are expected to serve, in a unidirectional manner, from a departure point to a destination.

Route directions

"When you get out of the subway station, you will be facing a movie theater. Proceed to the right and take a short subway passage. You reach a bank. Turn right after the bank. The museum is at the end of the street." Route directions are made of a sequence of statements of several types. Essential components are of course instructions that prescribe actions to execute, such as move forward, re-orientate, pass, skirt around, etc. Other statements are not prescriptive and aim primarily at describing aspects of the environment. They are intended to help an addressee create a more or less detailed representation of the to-be-traversed physical environment. These portions of discourse indicate the points of a path where critical actions are to be executed. Lastly, since this type of discourse is generally produced in a social context, with someone asking questions and someone providing answers, it contains statements with low informative content, but aimed at securing the addressee and reducing any feeling of uncertainty ("It is not too far," "You can't miss it").

The describer's task is to express verbally some form of spatial knowledge. If the describer has such knowledge available in an easily accessible form, the delicate aspect of the task will consist of conveying it through a sequence of adequate expressions. Deciding about the sequence of statements is not expected to create particular difficulties (contrary to descriptions of scenes, as reported in the previous chapter), but a selection has nevertheless to be made, in particular among the landmarks to be found along the route.

Route directions include a number of features that are also characteristic of route descriptions (as mentioned in the previous chapter). In particular, both are formulated from an egocentric perspective and use the language associated with that perspective. It is highly unlikely to hear route directions that are given from a survey point of view. Also, it is infrequent to find route directions introduced by an overview, with preliminary listing of the major parts of the route. The instant commitment to describing the route is probably due to the fact that, in its canonical version, the request for assistance is expressed on the spot and discourse is produced at once in the very environment where the first step will be made.

Delivering route directions takes place in various contexts and with various intents, with the consequence that verbal statements will be modulated by these contextual factors. Christoph Hölscher, Thora Tenbrink, and Jan Wiener (2011) proposed an insightful analysis of such factors by considering, beyond the classic case of directions provided to a foreign visitor, the situation where a person describes the route that he/she is about to engage in. Can we consider that the same cognitive act governs the generation of these two descriptions? Will the route described in both cases be the same? Will it correspond to the route effectively followed by the describer when he/she is invited to reach the same target from the same departure point?

When someone describes a route in response to a visitor's request, it is likely that the speaker selects a route that is both easy to follow and easy to memorize. In some cases, the speaker may be led to propose a route that includes a detour if a description is easier to process than that of a shorter path devoid of distinctive cues. But when the person is invited to describe the path that he/she will follow to reach the same target (a "prospective description" for one's own use), the directedness of the route may be sacrificed for the benefit of a shorter route.

In both scenarios (delivering a navigation plan for someone else or describing one's own plan), the route must be fully set out in order to be conveyed verbally. But when a person moves naturally towards a target, planning is likely to be implemented incrementally, during the execution. Here again, it is sound to consider that the underlying cognitive processes are not necessarily the same depending on whether someone actually performs travel or makes plans for a forthcoming trip. Behavioral analyses reveal that in "situated navigation," paths tend to be shorter (and possibly more complex) than those that are described. In addition, planning for someone else – the situation to which the present chapter is devoted – generates simpler (possibly longer) routes and more detailed descriptions than planning for oneself.

Models

Wolfgang Klein, from Nijmegen, is a linguist highly attentive to the pragmatic markers of human communication. He was the first to develop an analysis of interactive space-related situations, in particular those in which route directions are generated. In these situations, well known to sociolinguists, the speaker and the receiver must have the same (or sufficiently similar) representations of spatial concepts and the related lexicon. But we know that the meaning of spatial expressions conveyed by a particular statement depends on the positions of the speaker and the receiver. The referents of the adverb "here," of the preposition "behind" or of the locution "to the left of" depend on the place where the speaker stands while speaking, and thus on the circumstances in which discourse is generated. These "deictic" expressions must be interpreted in the particular spatial context of their production, by taking into account specific interaction schemas (see Klein, 1982; Wunderlich & Reinelt, 1982). These interactions can be interpreted in the broader context of the regulation of exchanges between a questioner and an answerer, which include counter-questions by which the answerer checks the state of knowledge of the questioner (see Golding, Graesser, & Hauselt, 1996).

Following sociolinguistics, cognitive geography has proposed models of the processes underlying the generation of route directions. In particular, Helen Couclelis, a geographer from Santa Barbara, proposed a model according to which an imagery-based representation of an environment can be "translated" into a linguistic form, but assuming that the process is mediated by a common underlying internal representation, an abstract "mental model," which is neither linguistic nor map-like. It is the information contained in this representation that is expressed in the form of verbal statements for the benefit of a questioner (see Couclelis, 1996).

A detailed model of the processes involved in the generation of route directions was proposed by the late Gary Allen (1997), from the University of South California. His analysis was based on a taxonomy of speech acts involved in direction-giving. His classification of the recurrent verbal devices used to give directions first distinguished those that define places, those which refer to landmarks, and those which describe movements to be executed. The most detailed part of Allen's analysis was about the nominals by which a direction-giver refers to relevant environmental cues, that is, landmarks (which are typically used as sub-goals along the route), pathways (which are the "channels of movement": streets, etc.), and choice points (such as crossroads, which afford several options with regard to pathways). Allen also devoted himself to the description of what he called "delimiters," that is, statements that designate distances or directions, or refer to places or objects serving as points of origin or destinations.

Based on these classifications, Allen delineated a set of rules for effective direction-giving, which he summarized in the form of the CORK (Comprehension of Route Knowledge) conventions. This set of principles was tested in field studies involving students of a university campus as participants (see Allen, 2000). The first principle pertains to the order in which statements are to be delivered. As expected, navigational performance is better when the order of presentation of directions is consistent with the spatio-temporal sequence of travel. The second principle is concerned with the reduction of uncertainty at points of the route where directional choices have to be made. At such places, performance is better when instructions are more detailed and refer more frequently to landmarks. The third principle has to do with the navigating person's preferences for particular spatial designations. For instance, it has been observed that in general men tend to privilege information about metric distances and compass direction, which suggests that this type of information might be of special value to them. On the other hand, women are known to prefer descriptions that emphasize environmental features, such as visual landmarks, which might incite direction-givers to use a spatial language adjusted to their expectations. Allen (2003) went on to investigate the role of gestures that accompany verbal route directions. The valuable feature of deictic gestures is that they map directly onto the environment shared by the questioner and the describer, at least at the starting point of the route. It is well established that gestures more generally play an important role in the expression and communication of spatial knowledge (see Alibali, 2005).

At about the same time as Allen introduced his CORK model, we developed thorough investigation of route directions, by focusing on the identification of their most crucial components to assist wayfinding (see Denis, 1997). We started with the assumption that when someone engages in the production of route directions to help an addressee, three cognitive macro-operations are implemented, each of which involving a set of more elementary components. The first operation consists of the direction-giver activating an internal representation of the territory in which travel will take place. The next operation consists in planning a route connecting the starting and the destination points within the relevant subspace of the activated representation. The last operation consists in formulating the procedure that the navigating person will have to execute to progress along the route and eventually reach the destination. Based on the analysis of a corpus of route directions collected on the Orsay campus and in urban environments, we developed a classification of the statements generated by describers.

At the first level of the analysis, the distinction was made between two families of verbal outputs. The first family includes statements that refer to landmarks, mainly three-dimensional objects which are natural parts of the environment (buildings, shops, monuments, etc.), as well as artifacts conveying directional information (signposts). The two-dimensional entities on which displacement is executed can also be used as landmarks (streets, squares, crossroads, etc.). The second family of outputs are instructions that prescribe which actions are to be executed (moving ahead or re-orientating). Most frequently, landmarks are mentioned to

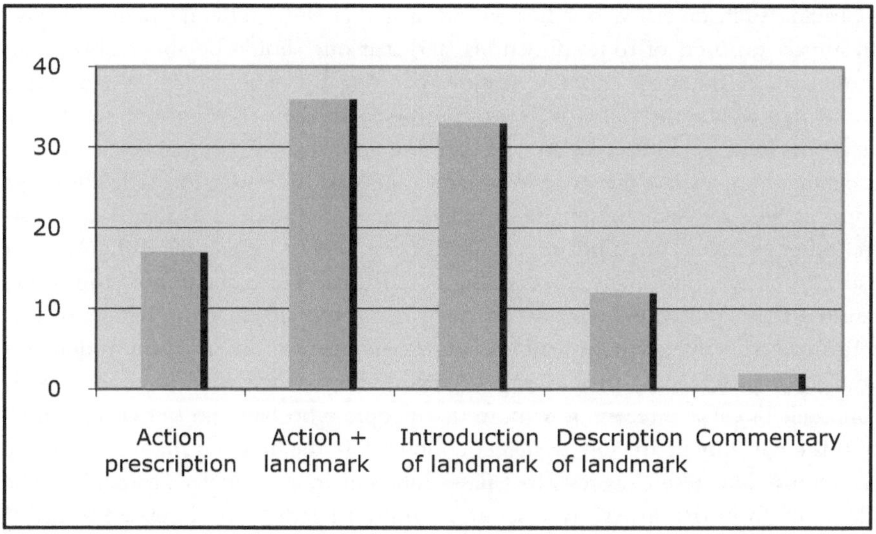

FIGURE 12.1 Frequencies of five classes of statements

signal the point where an orientation change has to be executed, but they may also serve other functions. This is the case when they have a confirmatory function during the carrying out of lengthy progressions without orientation change.

At a finer level, five classes of statements are defined, which appear with unequal frequencies, but are remarkably stable among various types of environments (Figure 12.1). It is noteworthy that more than 80 percent of statements are related to landmarks (the three classes at the center of the figure). This high density of descriptive information is to be emphasized especially if one considers that we are dealing with procedural discourse here, that is, discourse primarily intended to prescribe actions.

Validations

The most striking feature of a set of route directions provided by a sample of informants is the variety of the protocols, whether one considers their style, accuracy, length, or richness in detail. Admittedly, they all are organized according to a canonical structure which iterates a triplet of elementary actions: (re)orientation of the navigating person in a specified direction; engagement in progression; and reaching cues that mark the end of the traversed segment (the end of the last segment coincides with the destination point). But the diversity is remarkable. One finds very compact descriptions, concentrating on a limited number of prescriptions and referring to a minimal number of landmarks. To the opposite, one finds lengthy descriptions, resulting in highly detailed protocols, referring to abundant (even, overabundant) visual landmarks. From a describer to another, different sets of landmarks may be enumerated in route descriptions for the very same environment.

Facing such diversity, our first intuition is that these differences should have an impact on users of route directions, and that one should be able to bring out those that are the most effective in terms of navigational assistance. When judges are invited to rate the value of a set of protocols for their communicative value, a clear hierarchy is evidenced. At one extreme are protocols judged to clearly and accurately convey the pieces of information needed for a user to reach the target comfortably, and which contain all relevant (and only relevant) information. At the other extreme, one finds protocols that are likely to engender difficulties for users, by preventing them constructing an adequate representation of the route. Among these protocols, some suffer from being too succinct (therefore, of low informational value), whereas others contain an excess of information. Judgments are usually provided by people who have good knowledge of the described environment, but it is interesting to note that people who have no knowledge at all of this environment are able to detect contrastively the best and the poorest route directions. This result suggests that flaws inherent in a particular protocol can be identified from the analysis of discourse, without needing to be related to actual knowledge of the environment.[1]

Taking this diversity into account, a researcher may wish to identify the structure of what would be "ideal" route directions, the equivalent of a "macrostructure," which would adequately summarize the route and could be considered as its "leading thread." By applying simple statistics and also collecting responses from judges, a subset of statements can be isolated, which are interpreted as the necessary and just sufficient items to allow a user to perform easy navigation. This subset constitutes the "skeletal description," which only retains the essential of the whole set of individual protocols, with the smallest number of relevant landmarks and prescriptions to ensure efficient navigation. In short, the same "red thread" is followed by all describers, while being instantiated in qualitatively diverse discursive forms. The protocols assessed as the best ones by judges are those that have the largest overlap with the skeletal description.

Beyond these differences, descriptions of a given itinerary are governed by the same underlying structural organization. A route can be summarized as a sequence of segments which connect points where reorientations must be executed. It reflects a linear structure with a sequence of nodes connected by oriented segments. The same structure seems to guide people who provide route directions in the form of sketch maps. The approach developed by Barbara Tversky attests that verbal descriptions and schematic graphic figurations (or "depictions") contain, under different forms, the same basic information, suggesting that they are externalizations of a common underlying conceptual structure (see Tversky & Lee, 1998, 1999).

Once individual protocols of a corpus have been identified as "the best" and "the poorest" on the basis of judges' ratings, the ultimate sanction about their communicative quality and navigational assistance value consists in checking how they are effectively used on the spot by naïve navigators. In experiments designed to this effect, people are brought to an environment that is completely novel to them and receive route directions (generally in oral form). These directions are selected among those of the original corpus that have been rated either as the best or the poorest. The experimenter takes care that all are of similar length in order to avoid that potential differences in the navigation test are ascribed to differential memory load. The skeletal description is also submitted to participants in order to assess its value for navigational behavior. Such experiments were conducted in various environments, some very complex (such as the city of Venice; see Denis, Pazzaglia, Cornoldi, & Bertolo, 1999), some widely open (such as university campuses; see Daniel, Tom, Manghi, & Denis, 2003).

Volunteer navigators involved in these experiments exhibit quite contrasted behaviors. Measures used are the number of directional errors (taking a wrong street at an intersection), the number of pauses during navigation (reflecting the navigator's hesitation at a choice point), and the number of requests for assistance from the experimenter when the navigator experiences the feeling of being lost. All these indicators of difficulty are more marked for the navigators who have been exposed to the poorest descriptions than for those who received the good ones or the skeletal description. Navigation performed on the basis of a poor description also takes more time (even after deducting from total time the pauses as well as time taken to explore wrong segments). Furthermore, subjective assessment of traversed distances is overestimated by the participants who used a poor description. Lastly, the Venice study showed that when participants use a poor description, those who claim to prefer the survey perspective in spatial tasks commit more errors than those who prefer focus on visual memory of landmarks. On the whole, measurable properties of spatial discourse predict navigational performance that this discourse is supposed to serve.

Such attestations are in line with the criteria promoted by describers who aim at ensuring felicitous communication: a simple route, easy to identify and to follow, with as few direction changes as possible, with useful and distinctive landmarks (see Hölscher et al., 2011).[2]

Landmarks

Route directions are a domain where the significance of landmarks in spatial cognition is especially visible. The use of landmarks to identify an environment is essential for navigation. They provide means to localize and give an orientation to displacements. Stephen Hirtle, from the University of Pittsburgh, proposed a theoretical framework in which landmarks are conceived as being at the service of navigation by facilitating the identification of points of a path where navigational decisions must be made. In contrast with situations generating disorientation (absence of landmarks, uniform appearance of the environment, such as in mazes), landmarks contribute to cognitive differentiation of complex environments by offering distinctive cues that are easy to memorize. Singularity, contrast with background, and visibility are the characteristics that confer landmarks their special significance in the representation and description of an environment (see Sorrows & Hirtle, 1999).

Interestingly, people who have learned a route can demonstrate that they are still able to orient themselves when the landmarks at choice points are withdrawn (see Mallot & Gillner, 2000). But the value of landmarks in spatial learning cannot be underestimated, and route knowledge remains in general tightly associated with the memory of visual cues at choice points. Adding landmarks in an environment (be it natural or virtual) is beneficial to learning when this enrichment is ensured near the places requiring reorientation (see Heft, 1979; Jansen-Osmann, 2002).

We have noted that large individual differences exist as regards the frequency with which describers include landmarks in route directions. There are probably no means to determine what would be in the absolute the optimal number of landmarks to be mentioned in such discourse. Note that the critical components of route directions are not just landmarks, but the statements that *combine* reference to a landmark and specification of the action to accomplish when reaching this landmark. The most efficient directions are those where actions are tightly connected to landmarks, that is, those which closely link the descriptive and prescriptive parts of discourse. When describers are instructed to be concise, they shorten their discourse by lowering the number of landmarks and, as a priority, withdrawing those not associated with navigational decisions (see Daniel & Denis, 2004).

While landmarks may be found all along a described itinerary, their density is higher in the vicinity of the starting and destination points. In between, the places where reorientation will take place elicit more reference to landmarks, but such references are also found at intersections not requiring any reorientation. In other words, the places where faulty reorientations risk occurring are treated as critical nodes, requiring additional descriptive effort (Figure 12.2) (see Denis, Michon, & Tom, 2006; Ishikawa & Nakamura, 2012). These observations confirm the idea that landmarks play their cognitive role by signaling the parts of the route that require special attention. It remains, however, that a route which is rich in distinctive visual

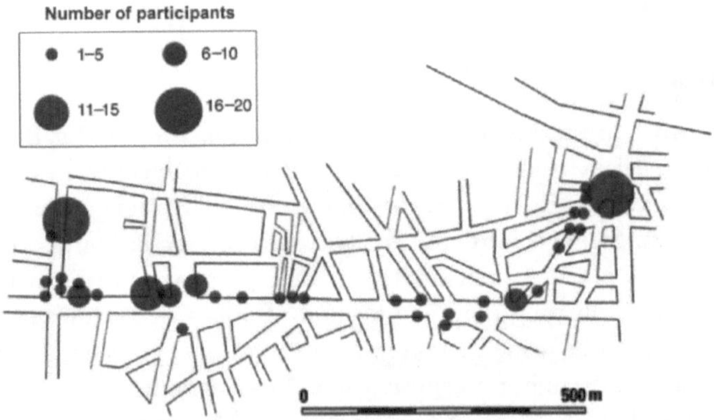

FIGURE 12.2 Number of participants who refer to landmarks along a route (from Place Saint-Michel to the Faculty of Medicine, in Paris). Reprinted from the chapter by P.-E. Michon and M. Denis in *Spatial Information Theory: Foundations of Geographic Information Science* (2001), by permission of Springer

cues – both at strategic points and along the paths – is better memorized than a route with fewer landmarks (see Denis, Mores, Gras, Gyselinck, & Daniel, 2014).

Not only do most describers spontaneously include references to visual landmarks, but users of route directions attest that they appreciate them. The privileged cognitive status of landmarks is due to the fact that they enable their users to build an advanced *visual model* of the critical aspects of the environment, which prepares them to make the relevant choices in those uncertain or ambiguous parts of the environment. Information conveyed by landmarks may even be preferred to information about pathways. Walkers perform better when they have received instructions about the landmarks they will encounter than when they only have a list of the streets to take. Furthermore, the processing of route directions focusing exclusively on streets is costlier and less favorable to memorization than when directions capitalize on landmarks. This effect may be explained by the special "vividness" or "distinctiveness" of information attached to landmarks (see Tom & Denis, 2003, 2004; Tom & Tversky, 2012).

Other functions of landmarks must be pointed out. We have essentially considered them as cues associated with a place and triggering actions ("At the church, turn left"). Each new sensory signal identified as relevant acts as a stimulus for directional decision. Landmarks may also have a function of confirmation or reassurance. It is the case when they are mentioned during the progression along lengthy segments, in order to confirm to people that they are still on the right track ("Follow the same road for about 500 meters; you will pass a movie theater, then a garage; when you reach the crossroads, turn right"). The landmarks here confirm to people that they have not deviated from the prescribed path.[3]

Other landmarks function as beacons. These are remote landmarks located in the neighborhood of the target. Beacons are directly connected to actions to execute on the path, but when heading towards such landmarks, gradual approach of the destination is guaranteed ("Notice the tower and walk in this direction"). There is experimental evidence of the efficiency of beacon landmarks in route learning, and sometimes their superiority relative to landmarks as mere associative cues (see Waller & Lippa, 2007; Wang, Mou, & Sun, 2014). While the effect is well demonstrated in experimental conditions designed for this purpose, remote beacons are not always available in natural environments.

Mental imagery

The role of landmarks in memory of spatial descriptions can hardly be mentioned without pointing to the role of mental imagery in the representation of space, in particular in the production and comprehension of route directions.

In tasks inviting people to provide route directions, Vanetti and Allen (1988) showed that participants with higher visuo-spatial imagery capacities generated more effective spatial discourse than low imagers. In our own analysis of route directions, we noticed that texts generated by people with the highest visuo-spatial capacities contained more frequent references to landmarks than those having lower capacities. This finding is in line with the assumption that if some people have greater proficiency at retrieving visuo-spatial information in memory, they are more likely to exploit such information and refer to it in their verbal productions (see Denis, 1997).

The relations between mental imagery and the representation of space have elicited a number of research programs (see Denis, Logie, Cornoldi, de Vega, & Engelkamp, 2001; Gyselinck & Pazzaglia, 2012). Differential approaches by psychologists of Paris and Padua aptly provide information on this matter. For instance, when people read a text made of a sequence of directional instructions within an environment ("Turn right to the picnic area," "Go straight to the playground," "On your left, you can see a lake," etc.), they later perform better in verifying spatial information if they have been invited to visualize the scenes while reading (in contrast to instructions to simply repeat statements). If one looks particularly at people with the lowest spatial capacities, the advantage that they enjoy from imagery instructions vanishes when they are involved in a concurrent task that taps their visuo-spatial resources. In contrast, people with the highest spatial capacities resist the effects of such interference (see Gyselinck, Meneghetti, De Beni, & Pazzaglia, 2009). The contribution of visual imagery to the creation of spatial models is acknowledged by a good number of researchers (although some arguments invite us to question the postulate of a functional role of imagery in this process; see Knauff, 2013).

If imagery associated with the processing of spatial instructions enhances performance in navigational tasks, it is appropriate to collect measures of individual imagery capacities in order to establish to what extent individual differences impact the processing and memory of spatial texts in the absence of any formal incitation to implement imagery. If imagery is brought into play spontaneously in the processing of route directions, people with high imagery capacities should outperform their counterparts in their processing and memorization of directions, especially if these are highly saturated with visual landmarks.

In our studies in Orsay, we investigated the conditions in which readers process the various components of route directions. For the whole set of participants, it appears that reference to landmarks ("You arrive at a telephone booth") requires longer processing time than prescription of associated actions ("Turn to the left of the booth"), which reflects the high degree of attention devoted to the objects that compose the visual scenes encountered along the route. After learning, landmarks are better recalled than actions. Furthermore, when landmark descriptions are enhanced with details ("The booth is accessible to wheelchairs"), their memorization is further increased. Lastly, participants classified as high imagers are faster at processing route directions than low imagers. This pattern of results confirms the well-established fact that imagery favors the processing of various forms of spatial discourse, and also suggests that people with a greater capacity to construct visuo-spatial images are in a better position to construct visuo-spatial representations based on the descriptive part of route directions. At the time of recall, both high and low imagers reach comparable performance level, but to attain this level, low imagers have to devote more time and effort than high imagers to constructing visuo-spatial representations that incorporate relevant spatial information (see Denis & Fernandez, 2013).

Children's route directions

Route directions have elicited special interest on the part of researchers in developmental psychology. Orientation in space is one of the most critical challenges met by young children in their daily lives. Research in this domain is to a large extent reliant on Siegel and White's (1975) model, which we presented in Chapter 5. Children in general are able to identify and use information provided by the objects met along a route. In line with Siegel and White's model, the youngest (7-year-olds) are more influenced by the presence, position and sequence of landmarks than older children (11-year-olds). For instance, they commit more reorientation errors when they are exposed to the route in which landmarks have been deleted. Lastly, the route maps drawn by the youngest do not reflect the integrated character of older children's maps. The latter, however, better recall landmarks that are the most relevant in assisting navigation than those of secondary importance (see Cohen & Schuepfer, 1980).

Of special interest are the studies in which children are considered in both their roles of speakers and receivers. The case of spatial instructions is taken as an illustrative case of communication situations where partners have to coordinate their interactions about a shared referent. Here, we consider the moves of a child in a room in search of a hidden object (by providing the child with a description of the path to reach the object). If the child is given a linear description of the steps to follow (as compared to a non-linear description), the search is more efficient (the child engages in fewer wrong trials). By doing this, the child shows early capacity (before 7 years) to benefit from the linearity that underlies a sequence of instructions. Then, it is interesting to involve the same child in a situation where he/she has to *generate* instructions to an addressee. The results show that these instructions adhere more and more to linear structure as age increases (from 4 to 8). Eight-year-old children delivering instructions to children of the same age are committed to linearity, but their effort is still more marked when their addressees are younger children. They appear to adapt to what they identify as younger children's limited cognitive abilities. By doing so, they demonstrate their capacity to adjust their communication style to their addressees, in such a way that their discourse offers more efficient assistance to the children who should take advantage of the linear structure (see Waller, 1986).

In other experimental contexts, the same participants play the roles of speaker and hearer alternatively in a task where one describes to the other a route across a map. Communication is via telephone, which forces the speaker to express verbally all the information that should be conveyed. Interesting are the efforts of speakers in order to distinguish among visually similar landmarks, but this strategy does not appear before the age of 10. At 7, children are minimally committed to that strategy. While 10-year-olds generate just a few more statements than the younger children, they adequately focus on distinctive features. At 10, speakers check whether hearers have understood their messages. In some cases, when the speaker's message is unclear, the hearer takes the initiative of controlling communication and eventually supervises the verbal interaction (see Lloyd, 1991).

It is around the same age (between 9 and 11) that children start giving attention to (and better memorize) the most "useful" landmarks along a route, such as those near choice points, thus relevant to reorientation (see Farran, Courbois, Van Herwegen, & Blades, 2012; Jansen-Osmann & Wiedenbauer, 2004). From age 10, landmarks to which a decision of reorientation is attached are more frequently mentioned than other landmarks (see Nys, Gyselinck, Orriols, & Hickmann, 2015). In Chapter 7, we pointed to the neuroimaging evidence that in adults the coding of these two sorts of landmarks is attested by distinct activations.

Notes

1 It is of special interest to note that the best protocols are provided by people identified as having high sense of orientation, as attested by self-rating questionnaires and visuo-spatial tests (Hund & Padgitt, 2010; Padgitt & Hund, 2012).

2 Another relevant factor is the compatibility between the speaker's and the user's spatial capacities. The success of "collaborative navigation" is affected by the quality of verbal communication between partners, but it is enhanced when both share a similarly high level of competence with respect to spatial orientation (He, Ishikawa, & Takemiya, 2015).

3 This function is particularly emphasized in the classification of landmarks proposed by Anacta, Schwering, Li, and Muenzer (2017).

PART V
Computation and technologies

On the issue of space, life sciences have good reasons to entertain close contact with computational sciences, given their capacities in the development of modeling and their support to technologies, in particular those dedicated to navigational assistance. These interactions make it possible to promote a unified conceptual framework intended to account for all cognitive agents confronted with spatial problems. Whether dealing with creatures equipped with a nervous system or machines holding algorithmic capacities in their software, the objective of cognitive science is to circumscribe their functional characteristics, including those common to natural and artificial agents.

Christian Freksa's work is to be mentioned here, as the promoter of an ambitious research program on spatial knowledge in Hamburg and Bremen, Germany. Freksa claimed the necessity of submitting natural and artificial cognitive systems to a common scientific approach, with the objective of identifying the principles that they share in the processing of spatial information. If such common principles were attested, this would create the need to investigate the capacity of these systems to communicate in efficient and profitable ways. This is an important issue since artificial intelligence models are frequently built on the basis of intuitions that are not always derived from empirical observations. If a cognitive function pertaining to space is to be implemented in an artificial device, the issue will be to establish whether it is relevant, interesting, or even indispensable, to ensure that it integrates what is known of its natural counterpart. Cognitively plausible computational models of human spatial reasoning will certainly be more valuable if they are adjusted to the capacities and expectations of their users (see Bhatt & Freksa, 2015; Freksa, 2004).

In this last part, special attention will be devoted to navigational aid systems, for which computer scientists and specialists of human–machine communication assign to automata the charge of conveying intelligible and effective spatial information to users. The advent of digital media (such as GPS-based navigational systems) offers new applications and new means for human cognition to deal with spatial problems. We will examine wayfinding situations where people resort to such systems. Here, the joint resources of geomatics and ergonomics are expected to contribute to the design of products that will deliver effective wayfinding assistance, attested by adequate behavioral indicators.

Lastly, we will look into virtual technologies, which open up new horizons for all scientists interested in spatial cognition (see Denis & Donikian, 2009; Péruch & Latini Corazzini, 2006). The visual scenes generated by virtual reality systems are not only available to perceptual exploration, but also to moves, as in immersive virtual environments. Virtual reality invites its users to entertain new relationships with fictional worlds. It is also a stimulus for new practices in society as well as in scientific research. In particular, it has amply modified practices in the handling of experiments, offering to researchers the possibility of having participants confronted with scenes that would be quite complicated and costly to set up with conventional means. It is the case when one has to simulate large-size or complex environments. But beyond the facilities offered by virtual reality, another aspect must be taken into account: the new space creates "affordances" and brings into play new forms of cognitive implication for users. This is of special importance when the created space is at odds with physical laws observed within what we are inclined to contrastively call, if giving a wink is permitted, "real reality."

13

SPACE AND COMPUTER SCIENCES

For lay people who use intelligent navigational aid systems, it is not always easy to distinguish that upstream of deliverance of any message of assistance, systems must be able to *represent* and *process* spatial information. At the origin of such performance are the empirical data collected for decades by psychologists and neuroscientists. These data have inspired computational models of space representation and reasoning. The models, in turn, are at the origin of systems intended to assist human displacement. The communities involved in these efforts remain close to each other in the development of their research programs (see Tenbrink, Wiener, & Claramunt, 2013).

The systems developed by artificial intelligence (AI) provide solutions likely to assist robot as well as human movements. According to a widely shared conception, formal systems rely on the same functioning principles as the cognitive system in charge of guiding human movements. In fact, in the spatial domain as in many others, computational systems frequently use algorithms that are not mere transfer prints of human capacities. Models issued from formal approaches are not necessarily pure replicas of human perception and reasoning. Their conception, however, obliges researchers to identify and understand the capacities that enable a human cognitive agent to represent space and move through it.

By maintaining as tight as possible a relation between spatial concepts used by humans and those that a formalizer uses to create artificial agents, researchers optimize the possibilities of communication between the two families of cognitive agents. By doing so, the researcher promotes navigational aid systems that complement (rather than replace) human cognitive competences (see Freksa, 2004).

The emergence of artificial intelligence models

Several modeling attempts appeared while AI tools were improving and their computational power was increasing. Two interesting conceptions, which today still have full heuristic value, were proposed by Kuipers (1978) and by Chown, Kaplan, and Kortenkamp (1995).

The TOUR model was developed by the end of the 1970s by Benjamin Kuipers at the Artificial Intelligence Laboratory of MIT. It is a *psychological* model of human knowledge about space. The original feature of the project is that the model is simulated by a computer program. The program takes as inputs simulated "observations" collected in an environment, assimilates them in a cognitive map, and uses these data to solve classic problems, such as route-finding or relative-position problems. Kuipers emphasized that TOUR is a model of human psychological functioning, whose simulation makes it possible to test the consistency of the postulated representations and inference rules. The author endorsed the cognitive map metaphor, whose properties (but not all) are similar to those of physical maps. A cognitive map is like a network made up of streets and intersections, even though the exact shapes and lengths of the streets are not always considered as essential. The map is also conceived as a catalog of routes, which are *procedures* for getting from one place to another.

The categories of spatial knowledge considered by Kuipers are the following: (a) routes, as ordered sequences of actions to be followed in order for someone to travel from one place to another; (b) the topological structure of the network, represented by descriptions of places and streets; (c) the relative position of pairs of places within a coordinate frame, which encodes orientation and distance information; (d) boundaries, which define regions to either side; (e) regions themselves, related to each other by containment. The representations used by TOUR include knowledge about particular environments, representations of a traveler's current position, and inference rules that allow TOUR to deliver adequate traveling instructions. A prescription of displacement can be summarized as a sequence of reorientation and progression instructions. The model is able to tolerate to some extent underspecified descriptions of a route or of the environment.

Later on, Kuipers developed a method that gives a mobile robot the possibility of exploring a previously unknown environment and creating a spatial map of it. The method differentiates itself from traditional approaches which essentially rely on metric information to describe an environment. It rests on three components: (1) procedural knowledge that guides the robot's moves, (2) a topological model of the environment, and (3) metrical descriptions that code the geometry of the environment. Building of a map by the robot takes place in spite of sensorimotor errors (whether random or systematic) that may happen during learning. The originality of the model is to exploit qualitative properties of explored spaces before geometrical properties, which are relatively prone to error (see Kuipers & Byun, 1991). By drawing on the properties of cognitive maps and promoting the concept of "qualitative navigation," this approach relies on a model of human spatial cognition and offers a method for guiding exploration and construction of maps by artificial agents (Kuipers, 2000).

Kuipers also developed a computational approach to a significant process, namely, the emergence of the subset of paths that form the "skeleton" of a cognitive map.

These paths are particularly rich in connections with other paths and in potential destinations. In planning travel, an expert wayfinder first looks for a route to the nearest point on the skeleton, then finds a route in the skeleton to a point near the destination, and finally follows the route that leads to it (see Kuipers, Tecuci, & Stankiewicz, 2003). The concept of skeleton within a cognitive map is reminiscent of the notion of "basic network" described by Pailhous (1970) in his analysis of the wayfinding behavior of Paris taxi drivers.

Another well-known computational model of spatial knowledge was proposed in 1995 by Eric Chown, Stephen Kaplan, and David Kortenkamp, from the University of Michigan. The model called PLAN is rooted in psychological and neuropsychological knowledge and illustrates the intent to treat navigation as an integral part of the general cognitive system rather than a system separated from the rest of cognition. The model refers to the neurobiological distinction between the ventral ("what") and the dorsal ("where") systems, with special attention given to the mechanisms of coding of the relative positions of objects in an environment. The authors develop their views within the then dominant conception of hierarchized acquisition of landmark, route, and lastly survey knowledge.

The original feature of the model is to rely on a system of representation of an environment as it is apprehended from a frontal view. The landmarks identified and memorized in the form of "prototypes" (constituent primitives of the spatial scene) are assembled in the form of ordered lists, which set up "topological routes." The model is able to grasp the content of a scene at a given time and to create "local maps" defined by the position and orientation of the mobile agent who is exploring the scene. However, these local maps do not stand with spatial operations applying to landmarks separated by long distances. This is why the model includes a further level, the "survey map" level, which ensures the positioning of particular landmarks in a broader frame of reference. Survey maps, however, do not lose the egocentric characteristic of local maps, given that they offer an oblique viewpoint from an elevated perspective from a particular point of the environment.

Extensions of PLAN were proposed later. In particular, Chown (2013) pursued the effort in order to connect the model to visual perception. His approach is based on the embodied cognition postulate, according to which representations created in a cognitive system are directly issued from perceptual interactions of an agent with the world, rather than deriving from codings that transform the representations into proposition-like abstract symbols. Chown adheres to a view of AI which is not massively focused on logic, statistics, and the manipulation of symbols. The concept of "spatial prototypes" is introduced to refer to primitives which spatial scenes are made of, including their dynamic properties (events, changes), as direct products of the processing of spatial information by the perceptual system. Another enrichment of the model resides in the consideration given to transitional places where a navigator passes from a visual scene to another. "Gateways" correspond to places of the environment where an occlusion is transformed in an opening, making it possible for a mobile agent to move from scene to scene. Passing through gateways results in significant changes in the agent's visual field.

Qualitative reasoning

When researchers commit themselves to the formal representation of spatial knowledge, they are confronted with the variety of situations in which such knowledge is used, and thus the variety of formal solutions likely to be implemented. For instance, quite different challenges are faced by researchers according to whether their objective is to create a detailed metric representation of a geographic area or to represent knowledge that will be used to describe places of an environment to assist wayfinding. In the first case, the formalizer needs to implement quantitative knowledge that will help to find absolute localizations of landmarks. In the other case, metric information drawn from the environment will not be that relevant, and therefore some amount of qualitative knowledge may suffice.

Christian Freksa (1991) promoted the view that a large number of situations calling for spatial reasoning pertain to the latter category, which consequently deserves AI researchers' attention. This form of reasoning, which only needs low-resolution representations to describe a state of the world, constitutes the domain of "qualitative reasoning." As a matter of fact, in empirical sciences, qualitative descriptions are considered inferior to quantitative ones. The privilege of the latter is due to the fact that a model that generates quantitative predictions is considered as being of greater value as compared to a model that only generates qualitative predictions. Furthermore, while qualitative information can be derived from quantitative information, the reverse is not true.

Here comes into play a conception that proved to be highly attractive in AI in the past, namely, the approach proposed by the computer scientist James Allen (1983) for the representation of temporal events based on qualitative information. Allen capitalized on the value of an approach aimed at representing the relative positions of events in terms of *intervals*, preferably to a representation referring to *instants*. He systematically reviewed the thirteen possible relationships – mutually exclusive – that two events unfolding over time may entertain. The temporal relationship between two events can be classified as one, and only one, of these thirteen cases (Figure 13.1). The approach developed to account for temporal information does not imply any measurement, hence its qualification in terms of *qualitative* approach.

The idea that emerged from this conception consisted of applying the formalism of time representation to spatial entities. Space shares with time the property of *extensivity*. As a result, the full set of topological relationships between regions taken two by two can easily be described. Two regions can be completely separate or be separate with a common boundary; or they can overlap, either completely or partially; or one can be comprised in the other, either with or without a common boundary. These topological relationships are relevant in the description of biological events, such as the successive phases of osmosis between two cells, or geographic entities, such as the relationships between territories. This approach has inspired qualitative formalisms and guided their application to various domains: planning, robotics, GIS (see Ligozat, 1993, 2012).

While geometry accounts for spatial relations by referring to quantitative measures, qualitative spatial reasoning infers spatial relations on the sole basis of qualitative concepts. A notion that takes on special relevance in this conceptualization is that of

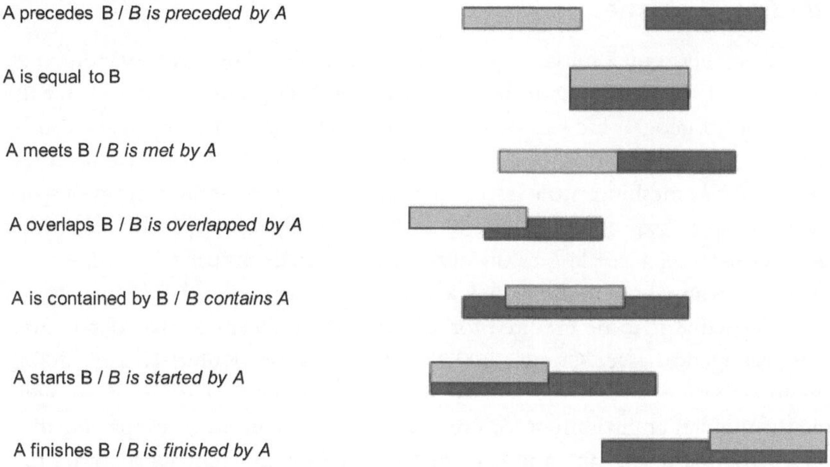

FIGURE 13.1 The thirteen qualitative relations between two temporal intervals. Adapted from C. Freksa (2013), by permission of the author

"neighborhood." The notion is particularly useful by contributing to the organization of elementary parts of an environment. The grouping of objects close to each other makes it possible to create hierarchies and unify neighbor objects in the form of higher-order sets. Beyond spatial neighborhood taken into account in the description of objects in space, it is appropriate to consider the notion of "conceptual neighborhood," which reveals the similarity (hence, the proximity) not only between *objects*, but also between *relations*. Topological relationships are thus distributed in a symbolic space that reflects their proximity (Figure 13.2) (see Freksa, 2013).

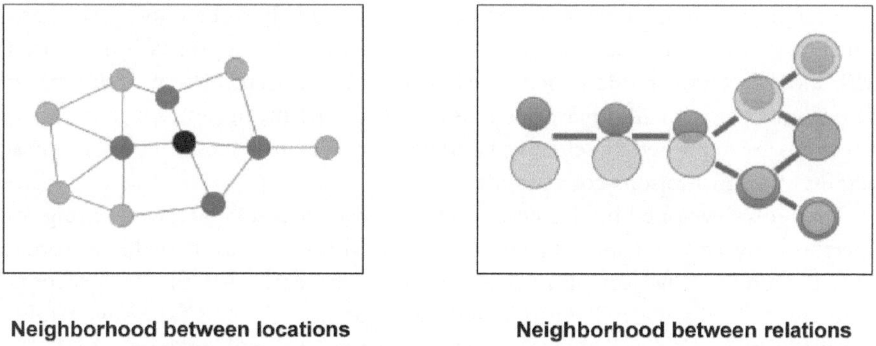

FIGURE 13.2 Graphic representation of neighborhood relations between locations (directly connected nodes represent spatial neighbors) and between spatial relations (directly connected nodes correspond to relations with minimal spatial differences). Adapted from C. Freksa (2013), by permission of the author

Wayfinding choremes

The notion of "choreme" initially appeared in geography. The term was forged by Roger Brunet (1980, 1987), with the intent to refer to elementary concepts for the representation of geographic spaces. For Brunet, the organization of space can be represented by a combination of a few models that have the status of conceptual primitives (such as meshing, attraction, or contact). These elementary types of space organization can be expressed graphically in the form of geometric schemas, which altogether constitute, according to the author, a "language of space."

Alexander Klippel, who launched a research program at the University of Bremen, Germany, is to be credited for adapting the concept to spatial situations involving navigation (see Klippel, 2003; Klippel, Tappe, Kulik, & Lee, 2005). Wayfinding choremes are expected to provide a model of actions of an agent moving through an environment (of urban type). They constitute a representation vocabulary that describes operations to be executed by an agent who comes to a decision point, mainly turning and non-turning actions. According to Klippel's conception of decision points, the fact of not taking a turn (e.g. proceeding forward at an intersection) is considered a decision, just as the decision of turning right or left when reaching an intersection. Wayfinding choremes are governed by a grammar that is modality-independent, but provides a basis for graphical, verbal or gestural externalizations. This grammar comprises mechanisms of re-expression which allow, for instance, elements of the same type to be chunked together. This is the situation that natural language manages in an instruction like: "Turn left at the third intersection," more appropriate than the repetitive instruction: "Pass the next intersection, pass the next intersection, turn left at the next intersection."

The chorematic approach draws on models of qualitative spatial reasoning by dividing space around a given point into eight sectors. The bisecting line of each sector corresponds to an axis that can be followed by an agent while progressing (Figure 13.3). The number of reference directions amounts to seven, given that the option of turning around and going back is less likely to be chosen as a continuation of the route than any of the other branches. This approach is in contrast with most direction models used in AI, where the concept of step backwards is treated on a par with all other directions at a choice point. In sum, seven potential directions are expressed by seven wayfinding choremes, to which a set of canonical linguistic externalizations corresponds.

The tools developed in choreme modeling make it possible to write strings of operations by concatenation of choremes, but also by creation of higher-order descriptive units. The formalism also introduces choremes that take into account the presence or absence of landmarks at decision points, as well as expressive refinements adjusted to individual preferences for variants of spatial language.

The specific value of wayfinding choremes is to propose a formalism of representation abstract enough to model spatial knowledge at a conceptual level, while remaining open to particular forms of expression when instructions are delivered to users. Moreover, the approach sticks to a functional perspective since intersections

FIGURE 13.3 Potential directions for conceptualizing goal-oriented actions. Left: the eight vectors originating from the point where the branches of the intersection meet. Right: the seven reference directions considered from the wayfinder's perspective. Reproduced from the article by A. Klippel, H. Tappe, L. Kulik, and P. U. Lee (*Journal of Visual Languages and Computing*, 2005), by permission of A. Klippel and Elsevier

are not treated from a purely geometric point of view (as meeting points of physical segments), but by reference to the action to be implemented at decision points.

Robotic challenges

The domain of robotics is so wide that it is impossible to do justice to it in this chapter. At the least, we will point to the challenges that the community dedicated to autonomous mobile robots is facing. In the first AI models of space representation, robots were implemented in the form of "simulators" in two-dimensional represented environments (see Kuipers & Byun, 1991). The challenge takes on a bigger dimension when researchers aim at building mobile artificial agents moving in real spatial environments.

The objective of autonomous mobile robotics is to build physical systems that can move purposefully in real-world environments that have not been specifically engineered for the robot, and without any intervention of a human operator. Such an objective requires creating robots with as much flexibility and autonomy as possible, that is, not simply executing pre-programmed sequences of actions in constrained environments. Instead, they should be able to exhibit adapted behavior in novel environments and unexpected or uncertain situations. This is why, in order to make robots able to tackle uncertainty, researchers have come to abandon the idea of completely modeling the environment when designing a robot, but rather endow the robot with the capacity of building this model by itself online. Proposals have been made to provide robotic systems with as robust as possible behavioral modules (see Saffiotti, 1997).

It is essential for any agent moving in the world – human or robot – to rely on a model of it. This representation forms the basis of reasoning that will be developed in order to guide future actions. In robotics, this representation includes elements provided from the outset by the designer, to be enriched by information that the system will collect autonomously later. A truly autonomous robot must be endowed with learning capacities, in order to encode relevant spatial relations between objects of the environment, such as neighborhood relations, connections between places, or locations of obstacles (see Freksa, Moratz, & Barkowsky, 2000; Thrun, 2003).

It is the objective of *robot mapping* to enable a robot to build and maintain a model of its environment by relying on spatial information collected during its displacements. The dual objective of this collection is to allow the robot to be aware of its location online and enrich its cartography for further navigational episodes, in other words, perform *simultaneous localization and mapping*. The two processes are closely linked. Errors of localization and errors in the creation of the model impact each other (see Chatila, 2008). At the service of learning, the robot uses sensors (such as sonar sensors which detect obstacles and guarantee moves without collisions) and odometric instruments, which make it possible for the robot to record and keep track of its moves. A representation of the environment is constructed, a "map" that has the property of "resembling" the spatial structure of the environment (see Wallgrün, 2010; Wolter, 2008).

During the construction of the representation, several problems have to be solved. The first one responds to the necessity of keeping track of already explored parts and parts that are still to be visited in order for the territory to be fully covered. Planning of these visits should ideally be made by keeping travel times and distances at their minimum. The currently constructed representation must include information that makes it possible to decide which is the next portion of the environment to be visited.

Another problem lies with the integration of newly collected data in the course of map building, and in particular their correspondence. It is essential to check whether two observations collected by sensors do or do not correspond to the same physical reality. It is unlikely that an environment to be captured at once. The most frequent case is when several observations have to be combined. This is the sensitive step of data merging, by which representations of a given object viewed from different perspectives are combined into a single piece of information. This aspect is particularly delicate in situations where learning of the environment is based on shared information with other agents, for instance in multi-robot mapping, when several robots cooperate in visiting different regions of an environment in order to establish its cartography (see Burgard, Moors, Stachniss, & Schneider, 2005).

A critical decision in robotic mapping is the choice between metric and topological approaches. Metric maps used in a number of robotic systems integrate the geometric properties of the environment. They express spatial relationships between entities in an implicit manner, by providing the coordinates of each spatial object within an absolute coordinate system. They rely on fine-grained grids, which guarantee good resolution, but require costly computations. Of course, information in coordinate terms must be as exact as possible and protect the system against any risk of inconsistency. Topological maps, on their part, describe the connectivity between places of an environment. Compared to metric maps, they are of less fine grain. They express spatial relations in the form of explicit statements that a given relation exists between objects. According to this approach, it is possible to keep track of the distinction between relations that have been effectively perceived (and recognized as valid) and those which have been deduced from these relations by a mechanism of inference (see Thrun, 2003; Wallgrün, 2010).

As for human navigation, landmarks are of special importance. They provide an aid to navigation, but also play the role of organizing concepts of space, as anchoring points which facilitate apprehension and memory of an environment (see Golledge, 1999). They are thus exploited as objects of the environment to specify the location of other objects or regions. In robotics, it is common to use the term "landmark" to refer to any identifiable point of space. It is probably preferable to reserve the term for points that are useful to support decision-making or help the robot to orient in space. Artificial landmarks can be placed in space (for instance, bar codes), and distinguished from natural objects, which have to be explicitly identified to be conferred the function of landmarks (see Hirtle, 2008). In this respect, robotics is dependent on methods that aim at optimizing the extraction of appropriate landmarks among a set of potential candidates (Madsen & Andersen, 1998).

In the design of robot systems, methods are sometimes developed without taking into account the fact that the surrounding world is changing, even in the very course of the process of robot mapping. Real environments are dynamic. One understands the value of considering robots as systems inscribed in time, and which should be able to continuously update their maps during their entire lifetime.

Biomimetic implementations

An autonomous robot in a real environment provides a behavioral model of living beings involved in navigational tasks. In this respect, robotics contributes to the understanding of human behavior. Explanatory hypotheses accounting for behavioral data may thus be tested when they are implemented in a robotic system, by the observation of the system's behavior in real environments. Of course, researchers have other means of modeling human behavior, in particular by formal simulations as those mentioned at the outset of this chapter. However, even with the most fine-tuned simulation, it is practically impossible to anticipate and model the whole set of situations reflecting the complexity of natural environments. Besides, researchers engaged in simulation run the risk of undervaluing the necessity of specifying each step of their computations, for instance by omitting to specify how a particular piece of information has been acquired by the system. Such overlooking is not allowed when the objective is to implement all computational steps in a system whose functioning will be tested empirically. Lastly, when mechanisms are inscribed in a system, their robustness is immediately attested or invalidated when they are probed on the ground. Such arguments are brought forward by the promoters of robotic approaches and, in particular, the designers of bio-inspired robots.

Hanspeter Mallot, neurobiologist at the University of Tübingen actively engaged in this approach, developed a program aimed at implementing spatial behavior of living organisms in autonomous mobile robots. His biomimetic approach is inscribed in a framework that provides a common description of mechanisms involved in navigation behavior of biological organisms and robotic systems. The objective, then, is to mimic animal and human behavior through a technical system whose performance is available to measurement (see Mallot, 1997).

In this respect, it is essential for researchers to carefully circumscribe the behavior under study. First, a distinction is to be made between two classes of behavior. On the one hand, one considers "local navigation," which requires on the part of the navigating organism the recognition of only one location, that is, the goal. On the other hand, one deals with wayfinding, which implies the recognition of several places and the representation of their relations. These places are not accessible to local navigation and may be outside the current perceptual field when displacement is initialized.

Considering wayfinding specifically, three variants can be described. The first method corresponds to situations where the recognition of the starting point triggers a local navigation mechanism that leads the agent to a goal by chaining successive segments. Every time a goal is reached, its recognition triggers a new response. This behavior does not involve any planning since spatial knowledge is limited to the recognition of a location and the triggering of the very next action, with each segment being generated independently of the others. Another method is based on topological navigation, which is adaptive in the sense that each represented location can be used to orient oneself towards multiple goals. This case makes it possible to generate integrated routes, whose collection results in a topological representation of the environment. Lastly, the third variant is survey navigation, which relies on survey representation of the environment. All known places and their spatial relations are integrated and articulated in a common reference frame, where the spatial representation is accessible as a whole. The relation between any two represented locations can thus be inferred, even if it has not actually been experienced. The representation, here, provides information on the basis of which a variety of adapted strategies can be generated (shortcuts, detours, bypasses, etc.).

Local navigation, recognition-triggered responses, and topological navigation are found through the entire animal scale. On its part, navigation based on survey representation appears to be limited to vertebrates. Interestingly, it is this kind of representation that is favored by classic robotics, for which spatial knowledge is best represented in the form of integrated global maps. Note, however, that the particular feature of biomimetic approaches is to consider the capacities implemented at all levels of the hierarchy of spatial behaviors (see Franz & Mallot, 2000).

Among the variety of biomimetic approaches, interesting attempts have consisted in having a robot learn the associations established by an ant between cardinal directions and landmark configurations when returning to the nest (Gaussier &

Zrehen, 1995) or developing robotic implementation of a neurophysiological model of the hippocampus in the rat (Burgess, Donnett, Jeffery, & O'Keefe, 1997). More recently, a robotic system has been proposed in which higher navigation capacities are used on top of simple, but reliable mechanisms, ensuring performance typically associated with survey representations (Franz, Stürzl, Hübner, & Mallot, 2008).

Researchers who adopt biomimetic approaches are pursuing a dual objective: to formulate biological hypotheses that can be tested in a rigorous manner in the real world, and to discover new mechanisms likely to be included in future navigational systems.

14

ASSISTANCE

In the research community dedicated to navigational assistance, concern is sometimes expressed about the objectives of designers of these systems (see Winter, 2009). In fact, there is still a gap between the (inevitably limited) services currently available and the considerable range of tasks and situations to which human cognitive capacities may apply. The issue is to assess the researchers' engagement to build machines endowed with genuine "spatial intelligence" and able to answer questions on localization, including the two basic queries that summarize questioning about space: "where (is X)?" and "how (to go to X)?"

It is generally considered as preferable for the outputs of a system to be delivered to users in a manner that imitates human communication. This requirement makes it necessary to develop ergonomics of communication about space. This concern leads some designers to leave out reference to quantitative geometric distances to the benefit of reference to landmarks, which is in line with users' preference for qualitative descriptions. In order to deliver adequate instructions, a pragmatic measure of informational content is required, especially in order to avoid information excess or redundancy (see Frank, 2003).

Spatial syntax

In addition to the ergonomic objective associated with the design of assistive systems, computer scientists are engaged in the characterization and formalization of the very concept of space, with an effort to put adequate mathematical and computing instrumentation at the service of this approach. The aim is to objectively assess the structural and functional qualities of a spatial environment (or, on the contrary, its defects or shortcomings), by executing computations expected to reveal either the ease of use or the complexity of infrastructures in which travel will take place. The hypotheses built during this approach aim at delineating the notion of "spatial syntax," intended to characterize environments in both formal and quantitative terms. The "intelligibility" of spaces to which this investigation applies is then submitted to empirical assessment, which calls on observation of actual behavior of human navigators (see Bafna, 2003).

An essential feature of human engagement is that in presence of empty spaces, human beings set about establishing structures, creating boundaries, building edifices, all actions intended to provide a context for their activities. This endeavor transforms and organizes initially empty spaces into *environments*, which display particular appearances and fit expectations such as harmony of layout, balanced structure, and visual agreement. Of course, the organization of space also takes into account the ease of traveling through it. As emphasized by Bafna (2003), the space of inhabitation is *configured*, in such a way that "the demarcation of boundaries allows particular relationships of access or visibility to emerge among the component spaces, and this in turn generates probabilistic patterns of movement and encounter within the population being housed" (p. 18). Once the organization of space is completed, people do not simply occupy certain "portions of space," but they live in "places" with which they develop various forms of attachment. The artifacts created in an environment impact their users' perception and cognition.

In computer science, a specialized community is dedicated to the description of spatial structures and the elaboration of concepts and tools intended to optimize the organization of human space. The same tools are also used profitably by designers of spatial forms. Humans, as agents of change of their own environment, have at their disposal computational resources to enhance their spatial creativity (see Bhatt, Schultz, & Freksa, 2013).

Attempts to specify the rules of a spatial syntax have developed rapidly. They make it possible to "create" spaces by taking into account empirically testable criteria and principles. Many examples are available in the fields of architecture and urban planning, resulting in the development of assisting devices for architectural design. Examples are available in various other domains, for which researchers have developed a range of tools.

To take one such example, Raubal and Egenhofer (1998) developed a computational method to assess the complexity of moves in large-size built environments like airports. In their approach to wayfinding, two critical elements were considered: places (decision points) where people have the opportunity to select among different paths, and clues (properties of the built environment) associated with these places. The method consisted in collecting frequent travelers' descriptions of the sequence of actions from the departure hall (the check-in counter) to a specific gate. Color slides of the environment were shown to elicit participants' descriptions. Their comments were recorded, with special interest for the description of difficulties encountered at decision points. Natural language protocols were rewritten in propositional form, with predicates expressing basic spatial relations ("schemata," in the authors' terms). For instance, the transcript "I move to the ticket counter" is re-expressed as "PATH (I, ticketCounter)." This formalism provides a standard structure to code critical elements of discourse and detect uncertainties expressed when the traveler reaches decision points. The method was used by Raubal and Egenhofer to compare descriptions collected in several international airports and thus identify the qualities and defects of their respective spatial arrangements. Their data show for instance that Frankfurt International Airport (generally considered difficult to navigate by its users) includes a larger number of decision points than Vienna International Airport (perceived as easy to navigate). Furthermore, when travelers reach decision points, the cases where clues are lacking or difficult to use are more frequent in Frankfurt than in Vienna. Such indirect measures of an environment's complexity thus appear to be correlated with travelers' subjective wayfinding experience.

Space of museums and art exhibitions has also attracted a lot of attention. The objective, here, is to establish whether the layout of museums affects the way visitors move through them and get into contact with displayed objects. The hypothesis is that visitors' behavioral schemas are connected to characteristics of an environment, such as visibility and accessibility. This expectation is relevant for "open spaces," which do not place objective constraints on visitors' movements. In this analysis, the first spatial units to be considered are "visibility polygons" (or "isovists"). A visibility polygon is the geometric structure that covers all the points that can be related to a root point by a visibility line that is not interrupted by any obstacle. When drawn on the floor, isovists capture the properties of the environment that affect movement. This is the way to define "accessibility polygons." Adequate software is used to generate these structures from any point of an environment. From these data, a set of "syntactic" measures is developed, resulting in indexes of connectivity and integration. These measures are considered as reflecting the "intelligibility" of the environment.

This analysis has been applied to museums and exhibitions by Wineman and Peponis (2010). While walking through a museum, visitors are implicitly guided along a network of primary paths, those linking to the largest possible number of destinations and having the highest degree of visibility and accessibility. The area of visibility and accessibility polygons can be computed, as well as "mean-depth value," a derived index defined as the distance between the root point of a polygon and the respective root points of all other polygons. The area of polygons appears to be correlated with behavioral measures that assess the attractivity of the exhibits, in particular the frequency of visitors' contacts with these objects. The phenomenon is confirmed when several exhibition settings are compared, with the same exhibits distributed according to different spatial arrangements. The accessibility of exhibits has a measurable effect on how museum space is explored.

Another type of environment has attracted the attention of specialists of spatial syntax, namely, libraries. Combined with the behavioral approach, spatial analysis can reveal objectively spatial areas where users are likely to be exposed to wayfinding problems. Li and Klippel (2012) compared different sections of a university library by relying on formal descriptions intended to quantify the intelligibility of corresponding spaces. Three components of spatial intelligibility are taken into account: visibility, layout complexity, and connectivity. Measures of visibility rely on visibility graph analysis, which provides quantified accounts of the size of visual access to the various parts of an area. Layout complexity is assessed by the computation of density (i.e. number) of direct connections between intersections (or decision points). Analyses reveal that the areas with higher layout complexity are also those with lower visibility. Lastly, connectivity is defined on the basis of lines drawn on axial maps, reflecting the set of potential paths within the environment. Higher connectivity of areas is correlated with higher visibility and lower complexity. When students are invited to locate books in different sections of the library with different spatial arrangements, the connections between these three properties are expressed behaviorally by shorter wayfinding times, shorter covered distances, and fewer detours. Spatial syntax methods thus prove to be effective by providing valid quantification of various aspects of the environment and to be predictive of the efficiency of human behavior.

Navigational aids

In recent decades, economically and technologically developed societies have become mobile information societies, thanks to the development of geolocation tools and services assisting people in decision-making about their displacements. Everybody now has access to systems endowed with GPS functionalities, which allow users to compute their position in an environment thanks to databases exploited by GISs. The existing systems deliver information in the form of instructions that use either verbal or map support. This technological and societal context engenders new needs, in particular the need for more proactive "engineering" of spatial cognition in order to assist system building and the development of services based on functional principles of human cognition and behavior. The fact that users are more and more involved in dynamic situations is further complicated by the capacity of geolocation systems, which can deliver online information to several users at the same time ("many-to-many communication"). Today's mobile geomatics is facing situations that it has engendered and did not exist only ten years ago (see Hirtle & Raubal, 2013; Montello & Raubal, 2013).

Early ergonomic approaches to navigational assistance were directed to vehicle drivers, with special concern for the optimal modalities of delivery of guidance information. On this subject, the value of vocal instructions was rapidly identified, by contrast with the use of maps, which solicit the visual system. In the design of verbal messages, principles of clarity were promoted, in a context where ambiguities must be avoided in order to maximize the chances of fast decision-making (see Riesbeck, 1980; Streeter, Vitello, & Wonsiewicz, 1985). Assistance to driving provided by embarked systems relied on cognitive models issued from ergonomic psychology, in particular through the identification of useful visual landmarks and the recognition of individual differences in driving strategies (see Baldwin, 2009; Burnett, Smith, & May, 2001).

The contribution of computational tools is especially appreciated when a navigational aid system is expected to define paths in response to specific individual requests, on the part of a pedestrian or a driver. Most navigation systems are designed to compute solutions based on the shortest path algorithm through a network of roads connected by nodes (see Dijkstra, 1959). However, one should also emphasize the value of the "simplest path" algorithm, that is, the path that would be altogether the easiest to describe, to understand, to memorize, and to execute (see Mark, 1985). The goal of this algorithm is to keep at a minimum the complexity of path description by controlling the amount of information required to negotiate each decision point, that is, the node where arcs connect in the network. The algorithm does not need any distance information, but only information about the complexity of instructions to be delivered at each node. Routes passing through nodes that involve the lowest number of theoretically possible reorientations will be favored, as well as routes with the overall lowest number of actual reorientations. Thus, routes avoid the most complicated intersections traversed by the shortest path, even if it is at the cost of increased total distance. This increase, in fact, is moderate (in the order of 10–15 percent on the average). The algorithm may thus be considered to be effective since it generates cognitively ergonomic instructions (see Duckham & Kulik, 2003; Richter & Duckham, 2008).

Another aspect of route directions calls for computational efforts to assist wayfinding, namely, the processing of visual landmarks. Their role is taken into account in computer modeling. A thorough review of landmarks and their function has been proposed by Kai-Florian Richter, from Zurich, and Stephan Winter, from Melbourne, two scientists highly committed to research in geomatics. Their synthesis of currently available data attests that there is still need to improve the quality of users' interactions with computerized navigational systems (see Richter & Winter, 2014).

In route directions, landmarks function as cues for triggering actions associated with crucial nodes of a path. We have seen the importance of speakers' efforts to extract these landmarks from the environment by taking their salience and visibility into account. For a designer, it is relevant to speculate about the possibility of automatizing this process. As a matter of fact, instructions delivered by most navigational aid services in urban environments rely on geometric information extracted from the street network, which is often the only available database. The consequence is that instructions are most frequently expressed in terms of distances and directions. Hence the idea of enriching route directions by adding references to local landmarks, as human describers do (see Basiri, Amirian, Winstantley, Marsh, Moore, & Gales, 2016).

Raubal and Winter (2002) developed a formal approach with the ambition of extracting visual landmarks from an environment by relying on measures of salience. The characteristics used to measure the contrast of a building relative to its neighbors are its façade area, shape, color, and visibility (from the egocentric point of view of the moving person). Statistical measures are used to assess the extent to which these values differ from those of adjacent buildings. The building with the highest salience measure in the vicinity of a decision point is selected and a verbal instruction is generated accordingly by the system. A computational model of extraction of the most salient buildings has been developed, whose evaluation

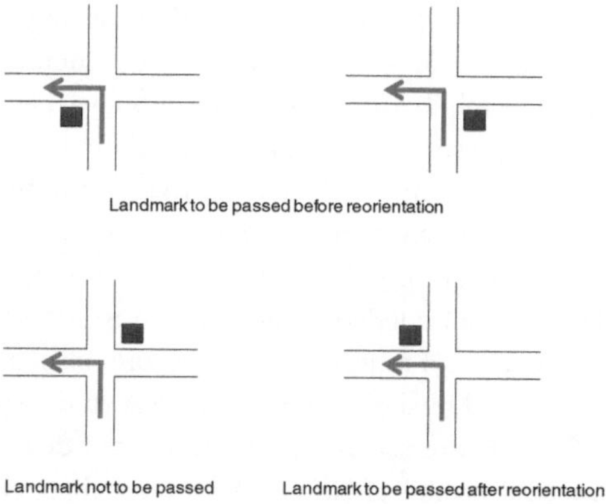

FIGURE 14.1 Different types of reorientation at an intersection as a function of the navigator's position relative to a landmark

attests for the good correlation of automatic landmark selection with on-site selection by human observers (see Nothegger, Winter, & Raubal, 2004). The system has also been enriched with a formal language based on choremes, which considers systematically all possible reorientations of a navigator relative to a landmark (Figure 14.1) (see Klippel & Winter, 2005). An alternative to this approach consists in operating automatic landmark extraction not from measures taken in the environment itself, but from data mining methods applied to spatial databases (see Elias, 2003; Tezuka & Tanaka, 2005).

Sadeghian and Kantardzic (2008) introduced further criteria for the automatic detection of landmarks. In particular, the determination of attributes of objects to be selected as landmarks could well extend beyond a building's fixed attributes, such as its height, width, and color, which may be of secondary importance. The idea, then, is to privilege attributes reflecting the frequency with which users interact with object candidates to the function of landmarks. For instance, a popular and well-attended building, like a museum, could be preferred to a visually salient building, but of low semantic value. The system could also aim at giving the status of landmark to entities other than buildings, including natural objects, which would be relevant for systems used in non-urban environments.

All these efforts at the crossroads of cognitive psychology and computer science have resulted in what seems to be the first system using landmarks selected on the basis of cognitive principles, the Whereis service, which covers all urban areas of Australia. Within each category (museums, theaters, restaurants, etc.), the system assigns weights to landmarks close to points of interest (primarily, decision points that imply a choice between several directions), then it annotates route instructions with landmarks based on the weighting system. Weights are adjusted to the mode of traveling, given that the visibility of landmarks is not the same for pedestrians and car drivers. This is a conceptual model that exploits easily accessible data, and is not necessarily biased by commercial priorities privileged by a service provider (see Duckham, Winter, & Robinson, 2010).

In natural assistive dialog, people who deliver information do their best in principle to adjust their message to the state of knowledge of their addressees. It would be appreciated if such "friendly" communication, which takes into consideration the characteristics of the questioner, could be implemented in navigational aid systems. However, most of them are based on the assumption of a questioner who is entirely ignorant about the surrounding environment. Customizing assistance requires assessment of the questioner's available knowledge and delivery of instructions adjusted to this knowledge. The problem is not an easy one, but it is amenable to formal approach and requires the development of algorithms intended to provide instructions guided by a dialog with the user of the system (see Richter, Tomko, & Winter, 2008). More recent efforts have been developed to automatically derive a mathematical model of landmark salience directly from descriptions given to a spoken dialog system by pedestrians walking around a university campus (see Götze & Boye, 2016).

By aiming at better adjustment of assistive messages, we must also take into account that some portions of a route traversing environments with varying degrees of complexity require more fine-grained descriptions than others. Humans demonstrate their capacity to switch from fine to coarse granularity. This is the case, for instance, in the description of "multimodal" travels (involving several modes of transportation). It may be necessary to enter into detail to describe the pedestrian or car access to a train station, but the describer will be more succinct about the rail portion of the journey. An artificial system endowed with this capacity is of great value for communication with users (see Tenbrink & Winter, 2009). The same applies for the differentiation (that humans achieve quite readily) between easy and more complex parts of a route. A navigational aid system should be able to devote special attention to the tricky parts of routes, for instance when the geometry of the street network does not fit the navigator's expectations. This is the case when a street makes a right-angle turn and keeps its name, whereas the street in direct linear continuation with the first one has another name. Automatic identification of tricky portions of a route and delivery of adequate descriptions by the system require special attention (see Hirtle, Richter, Srinivas, & Firth, 2010).

On the risks of cognitive disengagement

The benefits provided to users by spatial technologies are well recognized today. However, as with any technical innovation, these technologies must be examined for any possible adverse effects. The main concern in this respect is about the risk that systematic recourse to GPS-based navigational devices might contribute to decreasing users' orientation capacities inherent to their own perceptual and cognitive system. If the use of technologies in spatial decision-making has a valuable impact on navigational performance, it may also have less desirable consequences on learning and long-term mental representation of traversed spaces.

Empirical arguments confirm this worry in the context of car driving as well as pedestrian navigation. Let's first consider studies comparing a guiding system based on delivery of sequences of oral prescriptions of reorientation and a more classic situation where a driver drafts the route by studying a road map of the environment to be traversed. At the end of travel (in a driving simulator), the participants are tested in tasks expected to evidence spatial knowledge acquired (landmarks, route representation, survey representation). Drivers who have used the traditional method show better recognition of visual scenes of the environment and better memory of the order in which corresponding locations have been viewed. Sketch maps drawn by drivers who used guidance display scattered information and are sequentially fragmented. Of course, these findings do not mean that guiding instructions delivered online do not call for cognitive treatment on the part of drivers. Not only do participants have to process directions ("turn right"), but they also have to keep in working memory information about the proximity of forthcoming actions ("next turn, third exit"). They do not need to continuously maintain a sense of orientation throughout the journey, while the drivers of the other condition have to do so. The guiding system, in fact, has no favorable impact on the construction of a cognitive map. These results suggest the value of designing learning-oriented interfaces, as an alternative to navigational aids privileging minimizing uncertainty (see Burnett & Lee, 2005).

The same issue is encountered with computer-assisted pedestrian navigation. During a walking episode in a zoological garden, the users of a personal digital assistance computer acquire correct knowledge of traversed routes, but poor survey representation of the environment (whereas people performing map-based wayfinding show good route *and* survey knowledge). This outcome is not surprising if one considers that cognitive integration of information collected during navigation requires it to be encoded, transformed, and eventually stored in memory. These cognitive operations are not required from a system that essentially expects from users adequate locomotor reactivity to the delivered instructions. The cognitive processing of space is thus lowered to a minimum. The low degree of elaboration of representations collected during the journey can only result in weak spatial knowledge (see Münzer, Zimmer, Schwalm, Baus, & Aslan, 2006).

There exist ways to limit the dangers weighing upon the acquisition of spatial knowledge by users of guiding systems. A study of the representation of indoor space in a virtual four-story building shows the positive effects enjoyed by people who are prevented from passive monitoring of a navigation system. During navigational episodes, users may be offered the opportunity to continuously check their position on a map display included in the device. They are also invited to answer questions about their current position and orientation. Maintaining users actively involved with sustained mental effort results in better acquisition of spatial knowledge. As a matter of fact, in frequent occasions, such cognitive processing during a displacement is not really required, as is the case for occasional or recreational journeys. But this additional cognitive commitment may be recommended for operators whose professional duties make it necessary to acquire thorough knowledge of an environment (see Parush, Ahuvia-Pick, & Erev, 2007).

The fact remains that many signs converge to attest that when people have become familiar with an environment thanks to an assistive device, their later performance without this assistance is much poorer than for those who from the outset benefitted from maps or direct experience of the environment. In particular, in new tasks required from them, GPS users travel longer distances, make more stops, travel more slowly, make larger direction errors, and draw sketch maps with poorer topological accuracy, thus attesting for worse configurational understanding of the routes and the environment. Arguably, they eventually find their ways as other navigators do, but their navigation is less fluid and reflects poorer route knowledge (see Ishikawa, Fujiwara, Imai, & Okabe, 2008).

The development of spatial information technologies contributes to the strong tendency of superseding the use of our psychological abilities and routines by substitutive devices that ensure a good part of cognitive work. The threat of "technological infantilization," as harshly pointed out by Montello and Raubal (2013), is obviously a pitfall. GPS devices risk *substituting* themselves for human reasoning rather than helping people to *develop* it and increase its potential. The overgrowth of these systems without due concern about their cognitive impact may well, at a minimum, induct a degraded, in any case very limited, acquisition of spatial knowledge, with the triggering of a vicious circle that increases navigators' dependency on guiding systems, even for repeated displacements along the same route. In any event, the development of systems that would exclusively reinforce a navigator's egocentric perspective is not the most recommended formula. The systems that encourage survey perspective, and thus increase the possibility of building integrated spatial representations, should definitely be recommended (see Gardony, Brunyé, Mahoney, & Taylor, 2013; Richter, Dara-Abrams, & Raubal, 2010).

Anthropology itself is confronted with this challenge in its study of traditional societies, by distinguishing practices that "engage" individuals in their experience of the land and technologies that alleviate this engagement. To take a single example, Aporta and Higgs (2005) reported the changes that occurred in the life of Inuit hunters of the Arctic regions with the arrival of handheld GPS receivers. Their widespread availability had strong social and cultural effects. On the Igloolik Island, the Inuit people have mastered for centuries quite subtle knowledge that enables them to navigate through visually undifferentiated spaces. Such knowledge includes information like coastline shape, stone cairns, snowdrifts, wind direction, most of which are in fact practically ignored by GPS devices. In younger generations, quite open to technology innovation, hunters are worried by the potential risk of being deprived of any valid cues when the GPS receiver breaks down (due to very cold temperatures). GPS functionalities are evidently contributing to an extension of human capacities, but at the expense of losing effective knowledge of the territory by its inhabitants. Technology thus appears to reduce people's engagement with actual cognitive experience of the land.

Reliability?

The users of navigational aid systems are generally prepared to trust information delivered by these systems. Most of them do so. However, confidence in information provided may be questioned when a user detects divergences between transmitted instructions and the actually perceived environment. For instance, an instruction says: "Continue forward and turn left at the pharmacy." But the navigator, who sees a pharmacy in the distance, only sees a street on the right. Another situation may be that the navigator does find a left turn, but without an adjacent pharmacy. Which information should the walker rely upon? The landmark? Or the turn direction? Which decision to make when confronted with such inconsistency?

Tad Brunyé and his colleagues examined this kind of situation, where a navigator is in presence of incorrect, incomplete, or conflicting information, and starts wondering about the reliability of the information source. In an experiment, participants were invited to identify on a map the destination of a described route. In their effort to disambiguate instructions that contradict the configuration illustrated on the map, some systematically opt for the strategy that takes direction into account, while others exclusively rely on landmark information. For some people, there is no consistent reliance on landmarks or turn directions. The issue, then, is to know whether beyond any individual preferences, the option favoring one or the other piece of information is modulated by the *a priori* conception that people have of the reliability of the information source. According to whether information comes (or is said to come) from a GPS device or a human describer, will the user's approach be the same?

There is no doubt that both sources may be defective. But the interesting point is that different types of failure are spontaneously attributed to each by users. When instructions are reported to come from an automated system, the majority of potential navigators opt for the strategy that favors directional information. By doing so, they express greater reliance on what an assistive device communicates about directions. In contrast, when the source of information is a human person, they are more inclined to attribute inconsistency to turn-direction errors (possibly resulting from left/right confusion on the part of a human describer). Thus, a human is supposed to be less likely responsible for errors related to landmarks. The trust in either directions or landmarks is thus the result of users' different perception of the fallibility of the two sources (see Brunyé et al., 2015).

For the navigating person, the critical factor is the capacity to find out solutions when confronted with uncertainty created by contradictory pieces of information. Martin Tomko and Kai-Florian Richter, from the University of Zurich, have devoted an analysis of what they call "defensive wayfinding," by examining adverse situations in which a person develops special cognitive effort in order to correct or complement instructions that prove to be incompatible with the view offered by an environment. These situations may even include those where a user has to check for an alternative interpretation of misleading assistive messages just received. Tomko and Richter especially refer to ambiguous instructions, in which a critical element differentiating between several possible actions has been omitted by the describer (the wayfinder is invited to take "a street to the left," but there are in fact several streets on the left side of the intersection). They also consider over-specifications which may clutter up spatial descriptions, if only because an excess of information about the details of an environment increases the probability of including erroneous information, and consequently contradictions. Redundant information, acceptable in principle, becomes obviously undesirable if it is incorrect. The risk is that when navigators notice a discrepancy between what a system (or a human) has transmitted and the real configuration of the field, their confidence in the instructions just delivered is seriously affected. An analysis of these situations is required for the benefit of navigators placed in adverse conditions. Not only does it help us to understand how navigators eventually solve uncertain situations. It also inspires efforts to design modules enabling an automaton to follow flawed instructions provided by a human operator (see Tomko & Richter, 2015).

15
VIRTUAL SPACES

Virtual reality (VR) opens up new spaces, by offering us access to a three-dimensional sensory world intended to represent environments as realistically as possible, either environments that do exist somewhere else or fictional artifactual environments. Synthetic images that create these worlds are expected to render a visual appearance which elicits a feeling of immersion and presence in a veridical, geometrically coherent space. In every use of VR, whether it is ergonomic, recreational, artistic, museographical, instructional, therapeutic, even scientific, users are in the presence of a spatialized universe wherein they are invited to perform sensorimotor and cognitive activities that they have learned to master in the real world. It is expected from them to transfer to the new context their available repertoire of abilities, gestures, mental representations, and strategies.

With VR, the same people may be immersed in new environments, which require implementation of new gestural or cognitive competences. Such a situation occurs when intentional experimental distortions contradict expectations resulting from prior perceptual learning and expose people to unrealistic physical laws. This is the case, for instance, when a participant has to visually follow an object falling down, but that is not affected by gravity laws, or when a participant must execute manipulation or locomotion with force-feedback altered by the experimenter. All these situations warrant the effort of researchers not only to account for VR as an extension of the real world, but also as an experimental overture to new worlds, even improbable worlds.

Along with neuroimaging, VR is one of the technological innovations of the late twentieth century that have left the most significant imprint on cognitive psychology and neuroscience. Both have to do with space, although along different perspectives. Neuroimaging ensures visualization of the functioning of brain structures involved in cognitive activity. It reveals anatomical localizations and makes visible the signs of neuronal activity in these structures. The data collected contribute to the mapping of the brain and inspire models of the functioning of the nervous system and associated cognitive functions. On its part, the purpose of VR is to offer to human perception environments created by sophisticated digital systems. VR is a "space manufacturer." Both technologies were born in the same time window of the history of science and have come at the service of communities that had similar empirical needs. They are increasingly associated with the same research programs, as illustrated by many experimental examples cited in the previous chapters.

A tool for studying spatial representations and behaviors

In its most advanced form, space knowledge achieves an articulation between egocentric representations built during "first-person" locomotor experience and survey representations, which provide an allocentric view of an environment and correspond to the perspective typically associated with map reading. VR offers people an egocentric perspective and allows them to get acquainted with an environment by exploring it as they would for a real environment. A good illustration of this situation is available in a study by Francesca Pazzaglia and Holly Taylor (2007) (already mentioned in Chapter 11).

In this study, participants were submitted to the presentation of a virtual urban environment on a computer screen, which allowed them to discover the city through the equivalent of a locomotor exploration. They did so by following an avatar progressing along a predefined route. They were thus in the situation of acquiring route knowledge of the environment. Other participants were presented with a survey view of the same city on a screen, and were invited to watch a red dot moving along the same route. In both conditions, participants were instructed to pay attention to remarkable points (landmarks and intersections). Following the learning phase, participants were involved in tasks expected to assess the quality of the representation they had constructed. One task consisted in drawing a sketch map of the environment. Another was a classic pointing task (imagine that you are standing at a given location while facing another location, and point toward a third one). A third task was executed in the virtual environment, namely, retracing the previously learned route (by using arrow keys for moving forward, backward, right, or left).

The results showed that people who were assigned to the route condition performed with a lower error rate than those who learned the map. But the latter drew maps richer in information and provided more accurate pointing responses than virtual navigators. The pattern of results is in line with classic findings collected in navigation experiments in real environments. They are in accordance with the idea that information conveyed by VR generates a visual and cognitive experience which resembles what would be experienced in a real locomotor episode in a real city. Note that when VR is used by an experimenter as a substitute for a real situation, the implicit assumption is that the findings from the use of the virtual device can be likened to those resulting from a real ambulatory experience, in other words, that the cognitive processes implemented are comparable in the two situations.

Is it justified for a researcher to treat VR as a plain substitute for the real situations it simulates? The two situations are obviously distinct. If there are indications suggestive of quasi-equivalence, there may be other aspects inviting the researcher to be careful about any generalization. The answer to this question places demands on adequate experimental operations.

Validations

The comparative approach consists of examining spatial performance of people placed in real or virtual situations and evaluating the degree of similarity of their behavioral outcomes. To take an example, a classic approach consists of collecting directional pointing responses and distance estimations from participants who learned an environment either in real conditions or from a computer screen. Some experiments have attested that performance is not basically different in the two conditions (see Regian & Yadrick, 1994). This kind of result is taken as supporting the hypothesis that, at least as regards these particular tasks, knowledge of a configuration constructed from an artificial environment is functionally similar to what real-world experience makes it possible to build.

Quite accurate directional judgments can be made after virtual non-immersive exploration of a building (on a computer screen), but a measurable decline of performance appears when traversed routes involve a high number of turns (see Ruddle, Payne, & Jones, 1997, 1998). With technologies making people immersed in large-size virtual buildings (comprising more than seventy rooms), little difference is found in performance following learning with helmet-mounted or desktop displays, despite the differential impact of the two devices in terms of width of the visual field (see Ruddle, Payne, & Jones, 1999).

A comprehensive study has been devoted to the comparison of a large sample of experimental situations, in which participants had to estimate the direction of non-visible landmarks in a large-size environment. Directional estimates were made in one of five conditions: (a) in the real environment; (b) in an immersive virtual environment (with head-mounted display); (c) in a virtual desktop environment; (d) using a static panoramic image of the environment; and (e) using the traditional paper and pencil technique. Direction estimates were overall more accurate in the first three conditions than in the other two, and there was not much difference among the first three. One notes that kinesthetic and vestibular information available in the virtual condition with head-mounted display (but absent in the desktop condition) does not affect the accurateness of pointing responses (see Waller, Beall, & Loomis, 2004). Other studies have confirmed the quite small effect of proprioceptive information made available by head-mounted displays on spatial learning in a virtual maze (see Ruddle & Péruch, 2004).

A delicate matter is the possibility for the user of a virtual interface to have a veridical sense of distances, such that the structural organization of the virtual environment can be considered similar to that of the surrounding real world (see Waller, 1999). Although there is an explicit underlying metrics of technical devices that generate virtual images, the representation of geometric distances for the user is not an obvious matter. The estimation of distances in absolute metric units does not really make sense if the user does not dispose of a measurement scale. However, alternative methods have been envisaged. One consists of having participants execute two sequentially simulated self-motions and ask them which one covered a greater distance. Such distance discriminations prove to be accurate and reliable. Another approach consists in asking participants to estimate a traveled distance by using a reference distance available in the virtual environment. This method, used in CAVE immersive environments, reveals the good correlation between distances instantiated in the virtual world and participants' estimates, thus attesting consistent representation of traveled distances in virtual environments, even though some people tend to underestimate them systematically (see Frenz, Lappe, Kolesnik, & Bührmann, 2007).

Another approach to the similarity or differentiation between processes brought into play in virtual or real spaces consists of measuring transfer of learning between the two conditions. This provides a test of cognitive compatibility or equivalence between information provided by virtual and real worlds. In particular, will information acquired in a virtual environment be proficiently used in the real world? It is well established that after training with sufficient exposure to a virtual environment that simulates a real one, it is possible to transfer to the latter learning acquired from the former, and even to surpass performance of people who have been exposed to the real environment only (see Waller, Hunt, & Knapp, 1998). However, the advantage issued from this transfer is essentially evidenced when learning in the virtual environment has implied active navigational decisions on the part of the participants (see Farrell, Arnold, Pettifer, Adams, Graham, & MacManamon, 2003). Similar findings in the same line are obtained from people who have learned a large-size virtual environment (a scientific campus) and are later tested in the real-world environment, where they demonstrate accurate performance in direction and distance estimates (see Chabanne, Péruch, & Thinus-Blanc, 2003).

Differentiations

The quality of a VR device is to offer visual scenes that we know to be simulacra, but aim at achieving most closely the appearance and reactivity of the real world. The immaterial nature of the world that reaches our senses and the moderate richness of its visual appearance do not prevent this world conveying a genuine feeling of presence. These are two distinct experiences, of which one aims at becoming close to the other (which is its model), even though people keep in principle their capacity to be distanced from the virtual experience.

Beyond the structural similarities between real and virtual spaces, as attested by comparing behavior patterns in the two environments, there remains an irreducible cognitive difference between the two forms of experience. The question, here, is to establish whether what we learn from a virtual experience is fully comparable to what real experience brings to us. Research is committed to testing the limits of this alleged equivalence. Let's recall the study described above on the variety of modes for expressing spatial knowledge (Hegarty et al., 2006; see Chapter 4). Among the many tasks considered in that study, three concerned the representation of large-size environments. One consisted in learning a real environment, involving actual locomotor displacement guided by the experimenter. Another involved exploration of a virtual environment on a computer screen, also guided by the experimenter. In the third condition, the traversed environment was displayed via a videotape. The same participants were involved in all three learning conditions (for distinct, but structurally comparable environments). After learning, spatial knowledge was tested by direction and distance estimation tasks and recall of the environment in the form of a sketch map.

The results show that there are close inter-task and inter-condition correlations.[1] However, factor analysis clearly differentiates learning based on direct experience of the environment and learning from visual media. The contrast opposes, on the one hand, a situation which, in addition to vision, involves processing of vestibular and kinesthetic information and, on the other hand, situations with visual input mediated by a visualization device. Visual media, and especially the desktop presentation of virtual environments, offer visual scenes that only fill half of the visual field. As a consequence, objects present in a scene get out of the field more rapidly than in the real world. Working memory is under greater pressure since the internal representation of objects and their positions has to be maintained in an active state. Thus, there are distinct factors accounting for learning while moving in a real environment and learning from a VR experience.

Experience in real space? Experience in a virtual environment? Is it not relevant here to consider that beyond the similarity of cognitive activity elicited by both situations, this activity leaves different imprints on the brain infrastructure that supports it? A neuroimaging experiment by Emmanuel Mellet and his colleagues answers this question. The participants were invited to discover an environment located in a research building. For those of the first group, exploration was made by walking through a network of real corridors along which objects were posited as landmarks. The participants of the other group performed similar exploration while being immersed in the virtual version of the same environment containing the same objects. They followed the same virtual route while using a joystick to simulate locomotion (progression and reorientations).

Learning was tested the next day with a distance comparison task, which is known to reflect the quality of the topological representation of a learned environment. In this task, the two learning conditions resulted in identical performance (same rate of correct responses, similar response times). This similarity of performance was confirmed by the equivalence of sketch maps drawn by participants in both groups, who correctly recalled landmark positions. All this invites us to consider that both learning conditions generated similar mental representations.

However, further information was provided by the fMRI recordings collected during the distance comparison task (during which topological representations were activated). The results showed that a common bilateral parieto-frontal network (known to be involved in the processing of spatial representations) was activated in both groups, but also more specific activations. For participants who had learned the environment by direct experience, there was marked activation of the right premotor frontal region, which is known to be implied in locomotion. For participants who learned the virtual environment, there was an activation of a left parieto-frontal network, which is known to be involved in gestural control. The latter finding may be interpreted as being related to the use of the joystick in virtual learning. Thus, we are facing the case of two well-differentiated learning conditions that result in the construction of similar mental representations, but leave differentiated traces in the brain depending on the real or virtual nature of the learning environment (see Mellet, Laou, Petit, Zago, Mazoyer, & Tzourio-Mazoyer, 2010).

Although subtle, the differences between learning of real and virtual spaces have implications for cognitive scientists who intend to exploit the resources of VR. They invite researchers to be careful in their spontaneous tendency to generalize conclusions from findings collected in simulated environments to learning of real environments. Nevertheless, in spite of still limited realism of simulated environments and the absence of effective locomotor involvement of participants in most virtual devices, a clear belief is expressed for VR as a valid substitute for real spatial situations. Care and sense of nuance prove sometimes to become of secondary importance when they are balanced with the considerable opportunities offered by this tool.

Beyond the substitution model

A large area of research on spatial cognition nowadays relies on virtual technologies, without always being committed to the comparison between the real and the virtual. In these approaches, VR is used because it provides a robust methodology, which helps in attacking theoretical questions pertaining to the representation of space.

One of the most important contributions in this field is that of Hanspeter Mallot, already cited in Chapter 13 for his work in biomimetic robotics, and his colleagues at the University of Tübingen. His program is guided by the ambition to identify the minimal representation likely to account for the behavioral proficiency of an (animal or human) agent in an environment. The formulation of parsimonious explanations is frequently illustrated by specialists of insect navigation. Mallot proposed a minimalist theory of the cognitive map as the graph of recognized *views* of an environment and *movements* by which an agent passes from one view to another. These movements can be represented in the form of egocentric coordinates, without any reference to an external compass independent of the agent.

VR is put at the service of this approach in order to single out the cues used by an agent while building spatial knowledge. The value of this technology is to allow for high control of visual contents delivered to users, easy collection of behavioral data, and precise control of vestibular and proprioceptive information related to movements (Mallot's experiments are focused on visual processing). Experimental operations then consist in direct measures of behavior in a reasonably realistic environment. The environment designed in Tübingen is known as Hexatown (Figure 15.1). It is a virtual maze built on a hexagonal raster of streets where each decision point corresponds to the junction of three streets. All junctions have the same structure and look alike, except the three particular buildings located around each, which can be used as visual landmarks. This virtual world is displayed on a computer screen. The participants use a computer mouse to move about the environment by selecting sequences of "ballistic" movements (with only two possibilities: translation along a street until the next junction, and a 60-degree turn to the left or to the right). The participants have no visual access to any aerial view of Hexatown. The visual information is restricted to the view of the landmark they have just reached, and they do not pass any other distinctive landmark when progressing along the street to the next junction and next landmark.

The experiments using this device explore a variety of situations. Learning of the environment is based on the experience of movements, followed by tasks consisting of finding new routes, in particular the shortest route from a junction to another one in a remote part of the environment. As is classical in this paradigm, the error rate decreases in subsequent search tasks, and knowledge acquired along particular routes transfers to new routes, thus suggesting that goal-independent memory of space builds up in the form of "cognitive maps." When questioned about distances between pairs of views, participants provide more accurate estimates for route than for survey distances. Lastly, sketch maps drawn by good navigators are topologically correct and contain correct references to landmarks, while topology is not preserved by poor navigators, whose sketches include a restricted number of junctions and isolated landmarks (Figure 15.2).

FIGURE 15.1 Aerial view of Hexatown

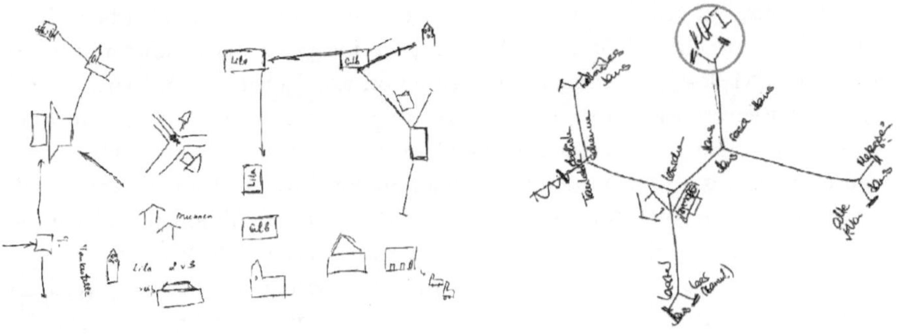

FIGURE 15.2 Sketch maps drawn by a poor navigator (left) and a good navigator (right). Reproduced by permission of H. A. Mallot

This approach confirms that spatial relations can be learned from active exploration of a virtual environment that imposes encoding in terms of *views* (accessible by motor commands). This is true, even under restricted viewing conditions (by creating a dark context, or restricting participants' field of view). Variations on the amount of visual information available have an impact, but the major lesson from these studies is the demonstration of transfer capacities based on *only local* views provided by the virtual device.

The fact that spatial information is acquired locally (by recognition of the reached view and the movement associated with it) attests that simple associations of this kind contribute to building genuinely configurational knowledge. Knowledge of the overall configuration of an environment experienced through local views can thus be acquired in a virtual environment.

These results collected with VR methods have a broader theoretical impact. The stereotyped strategy triggered by the recognition of particular views is an initial step, which opens up the possibility of planning new routes, then the acquisition of survey information. These competences form a hierarchy that reflects increased complexity, but also enrichment of spatial behavior repertoire in an evolutionary perspective. If hierarchized levels of complexity can be identified, the conception that is promoted is not that of levels that substitute for lower levels, but of their coexistence within the same navigational system, be it biological or artificial (see Gillner & Mallot, 1998; Mallot & Gillner, 2000).

There are other examples of creative use of VR to investigate cognitive mechanisms involved in route planning. In the framework of the theory of hierarchical encoding of space, the assumption of structuration of spatial memory in hierarchized levels is at the origin of interesting methodological extensions. According to this hypothesis, geographic entities can be grouped together and form superordinate entities in graph-like representations of space. Evidence for this conception is provided by experimental paradigms for which Tübingen researchers built a virtual environment made of a street network connecting places to which visual landmarks are associated. This environment is divided into regions that are made

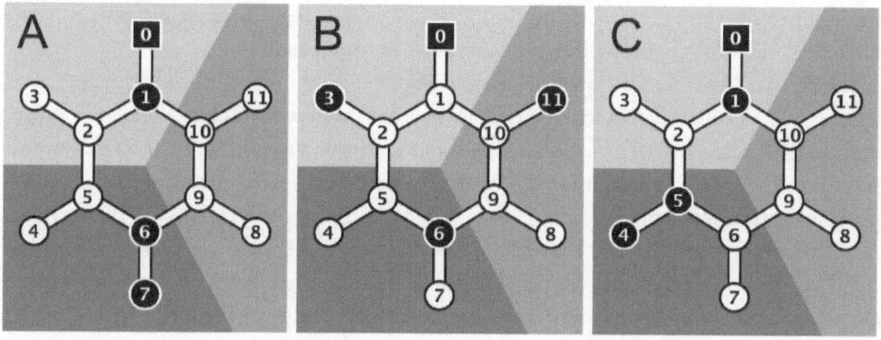

FIGURE 15.3 Routes crossing one, two, or three region boundaries. Reproduced from the article by J. M. Wiener and H. A. Mallot (*Spatial Cognition and Computation*, 2003), by permission of J. M. Wiener and Taylor & Francis

explicit to the users. After free exploration of the environment, the participants are invited to complete navigation tasks taking the shortest possible routes connecting two landmarks. In fact, several alternative routes of the same length are possible, but they imply crossing one, two, or three region boundaries (one boundary for routes from a region to an adjacent one, two or three boundaries for routes passing across one or two other regions) (Figure 15.3). The participants' choices reveal that they favor the routes that go directly to an adjacent region rather than routes of equal length passing through intermediary regions. By doing so, they demonstrate their preference for routes that cross fewer region boundaries. These data support the assumption that places can be grouped together with regions which form higher-level nodes in a graph-like representation of space. An inter-region spatial representation is then superimposed onto the inter-location organization within regions (see Wiener & Mallot, 2003).

Among the remarkable performances of researchers dedicated to VR, one should mention the attempts to test the capacity of non-human animal species to move through virtual environments. Such capacity has been demonstrated in chimpanzees and other non-human primates (see Leighty & Fragaszy, 2003). The efforts to create virtual systems for animals of smaller size, like rodents, have resulted in sophisticated technical systems. They attest that spatial learning capacity of rats immersed in virtual systems (adjusted to their morphology, visual system, and behavior) is comparable to that of the same animals in real environments (see Hölscher, Schnee, Dahmen, Setia, & Mallot, 2005).

Ergonomics of virtual interfaces

VR is a research tool, and also an *object* of research in full part. A specialized community deals with the properties of systems, their ergonomic qualities, and the effectiveness of their interfaces. Which parameters should be taken into account to ensure users' comfort and efficiency? Specific research programs are dedicated to the ergonomics of interfaces, the design and evaluation of users' interactions with virtual environments in a variety of domains of application.

Here, we consider the qualities of interfaces used to assess spatial learning. It is legitimate to expect that a virtual environment in which users collect all sensory information generated by locomotion helps them to move as easily as in the real world. In contrast, a purely visual interface based on desktop presentation does not convey proprioceptive information attached to the two main components of movement, namely, rotation and translation. Effective experience of both is nevertheless possible thanks to interfaces that make it possible for users to move effectively through a space that "contains" the virtual world and where they receive all proprioceptive information that they would collect in the real world. It is well known that locomotion strongly contributes to the feeling of presence in virtual environments. It also allows users to build better configurational representation of the traversed environment, which is attested by more accurate drawings of maps of the virtual room in which users have traveled (see Zanbaka, Lok, Babu, Ulinski, & Hodges, 2005).

In Cardiff, then in Leeds, Roy Ruddle developed a program of evaluation of interfaces at the service of investigation of human navigation. In these experiments, the participants' task is to move through a room-sized space, searching for targets randomly placed in several possible locations. This activity can be executed in one of four situations. (a) Vision: the only available information is visual, displayed on a screen. Moves in the environment (changes of position and orientation) are controlled by using a keyboard or a joystick. This situation is taken as a control condition relative to interfaces involving proprioceptive feedback. (b) Rotation: users access the virtual environment with a head-mounted display, while they remain in a stationary position. The only movement that they effectively perform is rotation when they reach a reorientation point. Progress through the environment is made with a joystick. (c) Translation: users, who also access the virtual environment with a head-mounted display, are standing on a linear treadmill and perform translation by actually walking, but their left or right reorientations are executed by using a joystick. (d) Walk: the participants wear a head-mounted display and move effectively in the environment (either freely or on an omni-directional treadmill), executing every required rotation and translation movement. As a result, they benefit from the proprioceptive and vestibular cues corresponding to the two components of movement.

When considering the number of targets effectively found in the virtual environment, it appears that the Walk condition results in much higher performance when compared with the Vision and the Rotation conditions. This means that the condition that combines translational and rotational information is the one where the most efficient navigation takes place. This superiority is confirmed even when all visual details of the virtual environment are removed (doors, furniture, objects, which normally help participants to maintain their orientation and keep in memory which parts of the environment have not yet been explored). Thus, even in a visually impoverished environment, the high performance level resulting from locomotion is maintained (see Ruddle & Lessels, 2009).

This research was extended to large-size environments, such as the VR tracking hall of the Max Planck Institute of Tübingen (which extends over more than 1,000 square meters). The results show that in the Walk condition, participants execute the most efficient displacement, with the shortest distances traversed. This condition, along with the Translation one, generates the most accurate direction and distance estimates. The experience of physical translation thus positively affects the formation of the mental representation of a large-size space, which is not the case with the sole proprioceptive experience of rotation (see Ruddle, Volkova, & Bülthoff, 2011).

It is noteworthy that the visual richness of information in a virtual environment is not required for effective navigation if full proprioceptive information is available. This is interesting if one considers the frequently repeated argument that a virtual environment requires a high level of visual richness and complexity to attain its full ecological validity. In fact, the concern is with the studies purportedly investigating navigation while using an interface restricted to a computer screen, a device that does not convey relevant proprioceptive information to users.

Note

1 Another study reported above (Chapter 8) found high correlation between spatial performance following learning of comparable real and virtual environments (see Cushman et al., 2008).

Epilogue

Spatial thinking

This book comes to an end. In writing it, I opted for giving special attention to what psychology and neuroscience tell us about the mechanisms serving spatial cognition. However, beyond my primary disciplinary affinity, I decided to explore the spectrum of the other disciplines whose aim is to understand space or implement suitable space-related practices.

By opening up the doors of other disciplines, we turn our minds towards multidisciplinary and interdisciplinary approaches. Metaphorically, exchanges between disciplines consist in experiencing *perspective changes*, a process thoroughly discussed in this book. Interdisciplinary exchanges allow researchers to confront views that are not aligned to each other from the outset, and ultimately take more than one point of view on their subjects.

Space, from all perspectives

When one reviews space and the range of its sizes (from small-scale figural space to large-scale geographical environments), the impression of lack of homogeneity is even enhanced by the variety of associated behaviors and modes of representations. The sense of fragmentation of the concept of space is exacerbated by the fact that its study calls for a variety of methods and measures. However, beyond that diversity, some federating concepts can be pointed out.

The most significant one is the reference to the *human* dimension of space. In line with Poincaré's and Merleau-Ponty's analyses, we emphasized that properly defining space implies reference to the individual's *body*, making it possible to give special relief to "representative space" (in contrast with "geometrical space"). Reference to the human is meaningful in both experimental and social sciences. We have seen that geographers are mindful of the fact that the establishment of sociality depends on people's shared experiences in shared environments.

Space cannot be reduced to physical areas and the geological matter making up the ground (see Lussault, 2007).

Other points of convergence are to be noted. All sciences dedicated to space and spatial knowledge, from psychology and neuroscience to language and computer sciences, are committed to the concept of *perspective* and to the distinction between the various points of view according to which an environment may be apprehended. Although egocentric and allocentric perspectives initially appear to be competing views on the world, their coordination is required for ensuring people's felicitous interactions with their environment. Computer science and human–machine communication do not ignore the indispensable coordination of these perspectives when engaged in the design of user-friendly navigational interfaces.

At this point of our journey, we are ready to delineate the concept of "spatial thinking," the state of mind that makes us look at the world and its actors by paying special attention to their spatial inscription, acknowledging in particular that we are inevitably *engaged* in environments and submitted to their constraints.

Spatial cognition is supported today by technologies, sometimes at the risk of the "technological infantilization" pinpointed by Montello and Raubal (2013), but also as factors of extension of profitable interactions with the world. Technologies should not make us forget or depreciate the power of spatial thinking and behavior in the absence of advanced forms of technological assistance. From navigation without instruments through Micronesian archipelagos, only supported by reading of the apparent movement of stars, until journeys through deserts over hundreds of kilometers by the Aboriginals of central Australia, ethnology provides examples of societies whose mastery of space ensures the survival of the group. By comparison with ordinary skills of the basic Western navigator, one is impressed by the considerable volumes of knowledge available to these expert navigators, the sharpness and sophistication of their computations, and the repeated attestation of their reliability (see Gell, 1985; Lewis, 1976).

However, beyond the complexity of situations, there does not seem to be any reason to postulate that essentially different cognitive functions are at work in these populations. The most impressive forms of spatial expertise are demonstrated by people whose general intellectual capacities remain otherwise in ordinary limits, but whose expert navigation abilities essentially result from sustained learning and intense practice (see Woollett, Spiers, & Maguire, 2009).

Virtues of spatial thinking

Thought takes a great diversity of forms, each of which is supposed to use its own system of representations and reasoning mechanisms. The essential feature of "spatial thinking" is to rely on the properties of space to organize and structure data or problems that we have to cope with, including problems that are not spatial in nature. Should this form of thinking reveal valuable properties, there would be good reasons to promote it, especially in educational contexts, and in many other occurrences of daily life (see Montello, Grossner, & Janelle, 2014; National Research Council, 2006).

Spatiality is a property of our world. Objects surrounding us have the property of extensivity. They occupy identifiable portions of space. Objects can then be defined not only by their intrinsic properties, but by a set of spatial features, such as their location, the distances that separate them, their possible overlapping, their connections, and their hierarchical organization (see Montello & Raubal, 2013).

Applying spatial thinking to spatial objects is quite common in our daily lives, such as comparing distances between points of interest. It also offers the possibility of switching to alternate perspectives and predicting how spatial configurations are affected when transformations are applied onto them. Developing spatial thinking about non-spatial objects can also make use of the distance concept, as an expression of similarity between elements distributed over a pluridimensional space. Spatiality offers a conceptual frame in which data of any sort can be integrated and organized into structured wholes. Spatial thinking then consists in expressing or re-expressing problems by using the language of space, which makes it possible to represent relationships among every sort of entity in terms of distance and build graphic figurations of them. Physical space then provides a metaphorical frame in which levels of similarity may be expressed through a gradient of proximity.

Metaphorical spatialization is implemented in the visualization of relations among abstract data, as in deductive reasoning. It is also used as a memory aid in the learning of complex materials. This is the case of the method of *loci*, developed by Simonides of Ceos at the time of Pre-Socratic Greece, or the method applied by Cicero to memorize parts of a discourse, associated with objects located along a familiar route. Creative use of visualization of spatial properties of complex objects has also proven to be productive in scientific discovery (see Miller, 1996). In sum, spatial thinking is endowed with great potentials due to its flexibility and transferability to a wide range of situations.

A special value of spatial thinking is to use intuitions that humans can mobilize without necessarily applying abstract reasoning. Admittedly, the propositional format is suitable to the description of spatial situations. We can, for instance, consider an abstract propositional rule stating that "objects that overlap must partially exist in the same space," but the reality expressed by these terms is in fact already inscribed in human experience and thus directly exploitable in spatial reasoning, with no need to rely on explicit engagement of this abstract knowledge (see Montello et al., 2014).

Here, we are dealing with quite general cognitive mechanisms, which transcend individual specificities of particular human communities. There are arguments in favor of the universality of mechanisms by which the human mind generates spatialized models (mental or physical maps). When people apprehend various aspects of a large-size environment from piecemeal sensory experiences, a valuable strategy for conveying that information to others consists in generating integrated pictorial or diagrammatic representations. This is the process by which people create maps by adopting a survey perspective. What they have to do for this purpose is to decide about appropriate scale reduction and keep control on the homogeneity of the representation. Remember that the oldest known map, found in the Neolithic site of Çatalhöyük in Anatolia, dates back to 6,200 BC, that is, long before the invention of writing. Prehistoric cartographic imagery, along with many ethnographic attestations, suggests that mapping behavior is a very old form of expression, which provides a beautiful example of a cultural universal (see Stea, Blaut, & Stephens, 1996).

The concept of spatial thinking is also attractive to AI researchers. It forces them to question the role of abstraction in spatial reasoning and to envisage the opportunity for inference engines to use representations of low abstraction level, as are spatial representations. For Christian Freksa, who has provided strong arguments in favor of qualitative reasoning (as mentioned in Chapter 13), physical space plays an essential role in cognition, not only as a domain where physical events take place, but also as a reference domain for interpreting non-spatial concepts. In particular, people's knowledge about the organization and laws of space (dimensionality, continuity, proximity, separation) serves as a natural cognitive interface between abstract knowledge and the real world. People can interpret non-spatial concepts by translating them into spatial concepts, which allows them to carry out computations in a "visualizable" domain and then repatriate the results into the original domain. For a general-purpose prover, each constraint to be taken into account must be made explicit, including physical laws which are part of permanently available knowledge in every mind (for instance, distinct solid objects never occupy the same physical space, or a physical object always occupies a single contiguous space). Recourse to spatial structures as support to reasoning in a concrete domain exempts the mind from costly abstract computational processes (see Freksa, 1991, 2013).

"There are no lost steps"

There are very few opportunities for humans to be in the presence of empty and inanimate space. In contrast with mathematicians' concept of space, our most common practice and experience is that of traveling spaces filled with objects and other people.

In an individual's history, the advent of locomotion marks a major transition, responsible for important psychological reorganizations. The effect of locomotion is to profoundly change our perceptual world and our mode of acquisition of environmental knowledge. In this respect, locomotion is the precursor of the development of spatial thinking, long before the emergence of language.

"Travel broadens the mind." This is the eloquent title of an article in which Joseph Campos and his colleagues of Berkeley proposed an insightful analysis of the impact of locomotor experience and independent mobility as factors of developmental change (Campos, Anderson, Barbu-Roth, Hubbard, Hertenstein, & Witherington, 2000). Locomotion makes a break of the infant's symbiosis with mother. By eliciting autonomous moves, it creates new conditions for the child's socio-emotional development. It creates opportunities for apprehending physical and social environments from novel perspectives and searching for new social signs. Locomotor experience also provides information that helps the child to distinguish between close and remote spaces and make new estimates of objects' sizes and distances. By traversing space, people encounter new affordances and encounter the Gibsonian evidence that "orientation is inseparable from locomotion" (Gibson, 1950, p. 230). The major significance of locomotion is to make investigative processes inevitable and elicit an exploratory frame of mind that will hopefully last a lifetime.

In accounting for how displacements help building spatial knowledge, there is much to learn from the strategies developed by blind people, for whom locomotion is an essential source of knowledge. When blind people have to learn the arrangement of a set of objects in a novel environment, their most efficient strategy is two-step. First, they make large exploratory cycles to comprehend the periphery of the room. Thereafter, they construct a representation of object-to-object relationships through back-and-forth moves between pairs of objects. Remarkably, independent of individual histories of visually impaired people (whether early or late blind), those who recourse to this two-step procedure are those who eventually perform the best. Overall learning of a configuration and its topological characteristics, followed by finer learning of distance and position relationships between objects, is an example of an especially effective differentiated locomotor pattern, provided it is practiced systematically over the whole area of concern (see Hill, Rieser, Hill, Hill, Halpin, & Halpin, 1993; Thinus-Blanc & Gaunet, 1997).

In locomotion, each step counts. Each step contributes to our knowledge of the environment and beyond, by incremental refreshing of visual scenes available to us. In short, locomotion (extended by mechanical devices that assist human travel) is an unlimited and permanent source of spatial knowledge, which reminds us of the revealing claim transcribed by André Breton: "Lost steps? But there's no such thing" (in *Nadja*, 1960).[1]

Eulogy of detour

Let's come to the end of our journey by considering a neglected, sometimes discredited, aspect of locomotor practice. Mastery of space, as measured by experimental records of human navigation, is generally expressed in terms of "finding the shortest path" or "traveling the quickest course." This bias reflects the implicit value conferred onto the achievement of "high performance" in spatial behavior. Wouldn't it be conceivable that understanding and mastery of an environment go through knowledge of space under all its facets by also including experiences along lengthy routes and sinuosities? Should we always aim at parsimonious models of traveling in terms of distance and time? Or, on the contrary, if circumstances allow, take our time along the route?

Strolling around is generally not associated with the idea of navigational performance, but rather with the motivation of contemplating scenes offered by an environment. Long paths, detours, circumnavigation, tacking, and winding all belong to a set of spatial behaviors that are hardly ever prescribed, but nevertheless highly valued by those who practice them. Setting oneself free from the compulsion of the shortest path and accepting the risk of getting lost (at least for a while, as the city of Venice offers unequaled opportunities) are conditions of singular navigational experiences. By strolling more than strictly necessary, a navigator enjoys occasions to scan an environment more widely, to identify what is useful (or unnecessary), relevant (or optional), and worthy of being memorized (or neglected). This approach creates a context for personal spatial experiences, more engaging than the plain objective of reaching the desired port by the shortest route.

There are very few studies devoted to this facet of spatial cognition. Researchers from the University of Alberta met the challenge and studied the behavior of children engaged in adventurous journeys in unfamiliar environments, with the potential risk of getting lost. The motivation of this research was to inform Civil Security Services in charge of protecting children from such dangers. The title of the article is quite explicit, by emphasizing the positive aspect of venturous locomotion: "Adventure as a stimulus for cognitive development" (see Cornell, Hadley, Sterling, Chan, & Boechler, 2001).

The starting point of this research is the observation that the familiar territory of children is generally limited. Children move through it first under their parents' guidance, then alone. Progressively, space across which they develop autonomous locomotion extends, which gives them the opportunity of testing navigational strategies to discover more of their environment. The experiment consisted in recording the behavior of children from a suburban neighborhood of Edmonton, aged 6 and 12, who were invited to start from their family home and explore freely their surroundings. Specifically, they were asked to walk until the farthest place that they had ever experienced (of course, they were escorted by an assistant). Not surprisingly, the results show that the limits of home range increase from younger to older children. Average traveled distances (as the crow flies) from the family home are 769 meters for the 6-year-olds and 2,501 meters for the 12-year-olds. The experiment also reveals that children's routes never follow the most direct line to the arrival point. The length of actually traversed routes systematically exceeds that of optimal paths. Sometimes, children take routes that they believe to be "shortcuts," but are in fact longer and more uncertain than the routes they already know.

The most striking finding of this study of adventurous navigation is that the ultimate point reached by children is beyond any expectation, especially for the younger. Before the experiment, parents had been interviewed and asked to state what was the farthest place the child had the opportunity of autonomously walking to from their home. While distances reported by parents of the older children are quite similar to those of actual journeys, they are largely underestimated by comparison with those of actual itineraries followed by the younger children. Six-year-old children, when reaching the end of their travel (769 meters from home), are well beyond the point declared as ultimately possible by their parents (507 meters), that is, a difference of about 50 percent. Another interesting indicator is the frequency of scanning behavior that accompanies walking (as measured by head rotations). The results show that 6-year-olds engage in active visual inspection of the traversed environment. Overall, while the younger children devote high visual attention to the environment, they show their capacity to navigate well beyond the limits that their parents think realistic, thus demonstrating unexpected adventurous engagement.

What is the value of adventurous course beyond a child's home range? Sense of freedom and excitement of discovery are enough to motivate a child to engage in exploratory navigation. Adventure also includes an experience of chance and risk, which contributes to entertaining a sense of wonder. What is true for children is likely to be the case at any age. Wandering is a good means to become aware of the closeness of places which have not been noticed during prior walking episodes, and a good means to entertain spatial thinking. Humans in motion are more likely to be inquisitive humans.

Richness of detours, value of drifting, benefit of wandering, and even of wrong turns? One of the great experts in signage and navigational aids, Per Mollerup (2013), from Melbourne, promotes the practice of "waylosing" (the counterpart of wayfinding) as a source of surprise and revelation, adding however that he does not advocate such practice on the way to the maternity ward, but reminding us that after all, America was discovered by navigators who had eventually lost their way.

Note

1 In addition to the enrichment of experience due to locomotion, we should remember the virtue of walking as the best way to put mind in movement, as recommended by Aristotle and the Peripatetic school!

BIBLIOGRAPHY

Afonso, A., Blum, A., Katz, B. F. G., Tarroux, P., Borst, G., & Denis, M. (2010). Structural properties of spatial representations in blind people: Scanning images constructed from haptic exploration or from locomotion in a 3-D audio virtual environment. *Memory and Cognition, 38*, 591–604.

Agel, H. (1978). *L'espace cinématographique* [Cinematographic space]. Paris: Jean-Pierre Delarge.

Aguirre, G. K., & D'Esposito, M. (1999). Topographical disorientation: A synthesis and taxonomy. *Brain, 122*, 1613–1628.

Aguirre, G. K., Detre, J. A., Alsop, D. C., & D'Esposito, M. (1996). The parahippocampus subserves topographical learning in man. *Cerebral Cortex, 6*, 823–829.

Alibali, M. W. (2005). Gesture in spatial cognition: Expressing, communicating, and thinking about spatial information. *Spatial Cognition and Computation, 5*, 307–331.

Allen, G. L. (1997). From knowledge to words to wayfinding: Issues in the production and comprehension of route directions. In S. C. Hirtle & A. U. Frank (Eds.), *Spatial information theory: A theoretical basis for GIS (Proceedings of COSIT 1997)* (pp. 363–372). Berlin: Springer.

Allen, G. L. (1999). Spatial abilities, cognitive maps, and wayfinding: Bases for individual differences in spatial cognition and behavior. In R. G. Golledge (Ed.), *Wayfinding behavior: Cognitive mapping and other spatial processes* (pp. 46–80). Baltimore, MD: Johns Hopkins University Press.

Allen, G. L. (2000). Principles and practices for communicating route knowledge. *Applied Cognitive Psychology, 14*, 333–359.

Allen, G. L. (2003). Gestures accompanying verbal route directions: Do they point to a new avenue for examining spatial representations? *Spatial Cognition and Computation, 3*, 259–268.

Allen, G. L., Kirasic, K. C., Dobson, S. H., Long, R. G., & Beck, S. (1996). Predicting environmental learning from spatial abilities: An indirect route. *Intelligence, 22*, 327–355.

Allen, G. L., Kirasic, K. C., Rashotte, M. A., & Haun, D. B. M. (2004). Aging and path integration skill: Kinesthetic and vestibular contributions to wayfinding. *Perception and Psychophysics, 66*, 170–179.

Allen, J. F. (1983). Maintaining knowledge about temporal intervals. *Communications of the ACM, 26*, 832–843.

Anacta, V. J. A., Schwering, A., Li, R., & Muenzer, S. (2017). Orientation information in wayfinding instructions: Evidences from human verbal and visual instructions. *GeoJournal, 82*, 567–583.

Annese, J., Schenker-Ahmed, N. M., Bartsch, H., Maechler, P., Sheh, C., Thomas, N., Kayano, J., Ghatan, A., Bresler, N., Frosch, M. P., Klaming, R., & Corkin, S. (2014). Postmortem examination of patient H.M.'s brain based on histological sectioning and digital 3D reconstruction. *Nature Communications, 5*: 3122.

Antonova, E., Parslow, D., Brammer, M., Dawson, G. R., Jackson, S. H. D., & Morris, R. G. (2009). Age-related neural activity during allocentric spatial memory. *Memory, 17*, 125–143.

Aporta, C., & Higgs, E. (2005). Satellite culture: Global positioning systems, Inuit wayfinding, and the need for a new account of technology. *Current Anthropology, 46*, 729–753.

Aradillas, E., Libon, D. J., & Schwartzman, R. J. (2011). Acute loss of spatial navigational skills in a case of a right posterior hippocampus stroke. *Journal of the Neurological Sciences, 308*, 144–146.

Aurnague, M., Vieu, L., & Borillo, A. (1997). La représentation formelle des concepts spatiaux dans la langue [Formal representation of spatial concepts in language]. In M. Denis (Ed.), *Langage et cognition spatiale* (pp. 69–102). Paris: Masson.

Avraamides, M. N. (2003). Spatial updating of environments described in texts. *Cognitive Psychology, 47*, 402–431.

Avraamides, M. N., Galati, A., Pazzaglia, F., Meneghetti, C., & Denis, M. (2013). Encoding and updating spatial information presented in narratives. *Quarterly Journal of Experimental Psychology, 66*, 642–670.

Avraamides, M. N., Loomis, J. M., Klatzky, R. L., & Golledge, R. G. (2004). Functional equivalence of spatial representations derived from vision and language: Evidence from allocentric judgments. *Journal of Experimental Psychology: Learning, Memory, and Cognition, 30*, 801–814.

Bachelard, G. (1964). *The poetics of space*. New York: Orion Press.

Baddeley, A. D. (2007). *Working memory, thought, and action*. Oxford: Oxford University Press.

Bafna, S. (2003). Space syntax: A brief introduction to its logic and analytical techniques. *Environment and Behavior, 35*, 17–29.

Bailly, F., & Longo, G. (2011). *Mathematics and the natural sciences: The physical singularity of life*. London: Imperial College Press.

Baldwin, C. L. (2009). Individual differences in navigational strategy: Implications for display design. *Theoretical Issues in Ergonomics Science, 10*, 443–458.

Barrash, J. (1994). Age-related decline in route learning ability. *Developmental Neuropsychology, 10*, 189–201.

Barrash, J., Damasio, H., Adolphs, R., & Tranel, D. (2000). The neuroanatomical correlates of route learning impairment. *Neuropsychologia, 38*, 820–836.

Bartolomeo, P., & Chokron, S. (2002). Orienting of attention in left unilateral neglect. *Neuroscience and Biobehavioral Reviews, 26*, 217–234.

Basiri, A., Amirian, P., Winstantley, A., Marsh, S., Moore, T., & Gales, G. (2016). Seamless pedestrian positioning and navigation using landmarks. *Journal of Navigation, 69*, 24–40.

Bellassen, V., Iglói, K., Cruz de Souza, L., Dubois, B., & Rondi-Reig, L. (2012). Temporal order memory assessed during spatiotemporal navigation as a behavioral cognitive marker for differential Alzheimer's disease diagnosis. *Journal of Neuroscience, 32*, 1942–1952.

Bergson, H. (1910). *Time and free will: An essay on the immediate data of consciousness*. London: George Allen & Unwin.

Berthoz, A. (2000). *The brain's sense of movement*. Cambridge, MA: Harvard University Press.

Beschin, N., Basso, A., & Della Sala, S. (2000). Perceiving left and imagining right: Dissociation in neglect. *Cortex, 36*, 401–414.

Beschin, N., Cocchini, G., Della Sala, S., & Logie, R. H. (1997). What the eyes perceive, the brain ignores: A case of pure unilateral representational neglect. *Cortex, 33*, 3–26.

Bhatt, M., & Freksa, C. (2015). Spatial computing for design: An artificial intelligence perspective. In J. S. Gero (Ed.), *Studying visual and spatial reasoning for design creativity* (pp. 109–127). Dordrecht, The Netherlands: Springer.

Bhatt, M., Schultz, C., & Freksa, C. (2013). The "*space*" in spatial assistance: Conception, formalization, and computation. In T. Tenbrink, J. Wiener, & C. Claramunt (Eds.), *Representing space in cognition: Interrelations of behaviour, language, and formal models* (pp. 171–214). Oxford: Oxford University Press.

Bird, C. M., & Burgess, N. (2008). The hippocampus and memory: Insights from spatial processing. *Nature Reviews Neuroscience, 9*, 182–194.

Bisiach, E., Brouchon, M., Poncet, M., & Rusconi, M. L. (1993). Unilateral neglect in route description. *Neuropsychologia, 31*, 1255–1262.

Bisiach, E., & Luzzatti, C. (1978). Unilateral neglect of representational space. *Cortex, 14*, 129–133.

Bloom, P., Peterson, M. A., Nadel, L., & Garrett, M. F. (Eds.) (1996). *Language and space*. Cambridge, MA: The MIT Press.

Bohbot, V. D., Iaria, G., & Petrides, M. (2004). Hippocampal function and spatial memory: Evidence from functional neuroimaging in healthy participants and performance of patients with medial temporal lobe resections. *Neuropsychology, 18*, 418–425.

Bohbot, V. D., Lerch, J., Thorndycraft, B., Iaria, G., & Zijdenbos, A. P. (2007). Gray matter differences correlate with spontaneous strategies in a human virtual navigation task. *Journal of Neuroscience, 27*, 10078–10083.

Bonnin, P. (Ed.) (2007). *Architecture: Espace pensé, espace vécu* [Architecture: Thought space, experienced space]. Paris: Editions Recherches.

Bourdon, B. (1902). *La perception visuelle de l'espace* [Visual perception of space]. Paris: Schleicher.

Bowerman, M. (1996). Learning how to structure space for language: A crosslinguistic perspective. In P. Bloom, M. A. Peterson, L. Nadel, & M. F. Garrett (Eds.), *Language and space* (pp. 385–436). Cambridge, MA: The MIT Press.

Brentano, F. (1988). *Philosophical investigations on space, time and the continuum*. London: Croom Helm Ltd.

Breton, A. (1960). *Nadja*. New York: Grove Press.

Brunet, R. (1980). La composition des modèles dans l'analyse spatiale [The composition of models in spatial analysis]. *L'Espace Géographique, 9*, 253–265.

Brunet, R. (1987). *La carte mode d'emploi* [Maps: How to use them]. Paris: Fayard/RECLUS.

Brunyé, T. T., Gagnon, S. A., Gardony, A. L., Gopal, N., Holmes, A., Taylor, H. A., & Tenbrink, T. (2015). Where did it come from, where do you go? Direction sources influence navigation decisions during spatial uncertainty. *Quarterly Journal of Experimental Psychology, 68*, 585–607.

Brunyé, T. T., Rapp, D. N., & Taylor, H. A. (2008). Representational flexibility and specificity following spatial descriptions of real-world environments. *Cognition, 108*, 418–443.

Brunyé, T. T., & Taylor, H. A. (2008). Working memory in developing and applying mental models from spatial descriptions. *Journal of Memory and Language, 58*, 701–729.

Bryant, D. J. (1997). Representing space in language and perception. *Mind and Language, 12*, 239–264.

Burgard, W., Moors, M., Stachniss, C., & Schneider, F. E. (2005). Coordinated multi-robot exploration. *IEEE Transactions on Robotics, 21*, 376–386.

Burgess, N. (2006). Spatial memory: How egocentric and allocentric combine. *Trends in Cognitive Sciences, 10*, 551–557.

Burgess, N., Donnett, J. G., Jeffery, K. J., & O'Keefe, J. (1997). Robotic and neuronal simulation of the hippocampus and rat navigation. *Philosophical Transactions of the Royal Society of London B, 352*, 1535–1543.

Burgess, N., Trinkler, I., King, J., Kennedy, A., & Cipolotti, L. (2006). Impaired allocentric spatial memory underlying topographical disorientation. *Reviews in the Neurosciences, 17*, 239–251.

Burnett, G. E., & Lee, K. (2005). The effect of vehicle navigation systems on the formation of cognitive maps. In G. Underwood (Ed.), *Traffic and transport psychology: Theory and application* (pp. 407–418). Amsterdam: Elsevier.

Burnett, G., Smith, D., & May, A. (2001). Supporting the navigation task: Characteristics of "good" landmarks. In M. A. Hanson (Ed.), *Contemporary Ergonomics 2001* (pp. 441–446). London: Taylor & Francis.

Buzzati, D. (1952). *The Tartar steppe*. London: Secker & Warburg.

Byrne, R. M. J., & Johnson-Laird, P. N. (1989). Spatial reasoning. *Journal of Memory and Language, 28*, 564–575.

Byrne, R. W. (1979). Memory for urban geography. *Quarterly Journal of Experimental Psychology, 31*, 147–154.

Campos, J. J., Anderson, D. I., Barbu-Roth, M. A., Hubbard, E. M., Hertenstein, M. J., & Witherington, D. (2000). Travel broadens the mind. *Infancy, 1*, 149–219.

Cantagallo, A., & Della Sala, S. (1998). Preserved insight in an artist with extrapersonal spatial neglect. *Cortex, 34*, 163–189.

Canter, D., & Tagg, S. K. (1975). Distance estimation in cities. *Environment and Behavior, 7*, 59–80.

Carbon, C.-C., & Leder, H. (2005). The Wall inside the brain: Overestimation of distances crossing the former Iron Curtain. *Psychonomic Bulletin and Review, 12*, 746–750.

Carreiras, M., & Codina, B. (1992). Spatial cognition of the blind and sighted: Visual and amodal hypotheses. *Current Psychology of Cognition, 12*, 51–78.

Cattaneo, Z., & Vecchi, T. (2011). *Blind vision: The neuroscience of visual impairment*. Cambridge, MA: The MIT Press.

Cattaneo, Z., Vecchi, T., Cornoldi, C., Mammarella, I., Bonino, D., Ricciardi, E., & Pietrini, P. (2008). Imagery and spatial processes in blindness and visual impairment. *Neuroscience and Biobehavioral Reviews, 32*, 1346–1360.

Cauvin, C. (1998). Des transformations cartographiques [On cartographic transformations]. *Mappemonde*, No. 49, 12–15.

Cauvin, C., Escobar, F., & Serradj, A. (Eds.) (2007–2008). *Cartographie thématique* [Thematic cartography] (5 vols.). Paris: Hermès/Lavoisier.

Cauvin, C., & Schneider, C. (1989). Cartographic transformations and the piezopleth maps method. *The Cartographic Journal, 26*, 96–104.

Chabanne, V., Péruch, P., & Thinus-Blanc, C. (2003). Virtual to real transfer of spatial learning in a complex environment: The role of path network and additional features. *Spatial Cognition and Computation, 3*, 43–59.

Chapelon, L., & Emsellem, K. (2008). L'interface: Contribution à l'analyse de l'espace géographique [Interface: A contribution to the analysis of geographic space]. *L'Espace Géographique, 37*, 193–207.

Chase, W. G. (1983). Spatial representations of taxi drivers. In D. Rogers & J. A. Sloboda (Eds.), *The acquisition of symbolic skills* (pp. 391–405). New York: Plenum.

Chatila, R. (2008). Robot mapping: An introduction. In M. E. Jefferies & W.-K.Yeap (Eds.), *Robotics and cognitive approaches to spatial mapping* (pp. 9–12). Berlin: Springer.

Chown, E. (2013). Spatial prototypes. In T. Tenbrink, J. Wiener, & C. Claramunt (Eds.), *Representing space in cognition: Interrelations of behaviour, language, and formal models* (pp. 97–114). Oxford: Oxford University Press.

Chown, E., Kaplan, S., & Kortenkamp, D. (1995). Prototypes, Location, and Associative Networks (PLAN): Towards a unified theory of cognitive mapping. *Cognitive Science, 19*, 1–51.

Ciaramelli, E. (2008). The role of ventromedial prefrontal cortex in navigation: A case of impaired wayfinding and rehabilitation. *Neuropsychologia, 46*, 2099–2105.

Cohen, R., & Schuepfer, T. (1980). The representation of landmarks and routes. *Child Development, 51*, 1065–1071.

Collett, M., & Collett, T. S. (2000). How do insects use path integration for their navigation? *Biological Cybernetics, 83*, 245–259.

Collignon, B. (2006). *Knowing places: The Inuinnait, landscapes, and the environment.* Edmonton, Canada: CCI Press.

Committeri, G., Galati, G., Paradis, A.-L., Pizzamiglio, L., Berthoz, A., & Le Bihan, D. (2004). Reference frames for spatial cognition: Different brain areas are involved in viewer-, object-, and landmark-centered judgments about object location. *Journal of Cognitive Neuroscience, 16*, 1517–1535.

Conty, P. (2002). *The genesis and geometry of the labyrinth: Architecture, hidden language, myths, and rituals.* Rochester, VT: Inner Traditions.

Corkin, S. (2002). What's new with the amnesic patient H.M.? *Nature Reviews Neuroscience, 3*, 153–160.

Cornell, E. H., & Greidanus, E. (2006). Path integration during a neighborhood walk. *Spatial Cognition and Computation, 6*, 203–234.

Cornell, E. H., Hadley, D. C., Sterling, T. M., Chan, M. A., & Boechler, P. (2001). Adventure as a stimulus for cognitive development. *Journal of Environmental Psychology, 21*, 219–231.

Couclelis, H. (1996). Verbal directions for way-finding: Space, cognition, and language. In J. Portugali (Ed.), *The construction of cognitive maps* (pp. 133–153). Dordrecht, The Netherlands: Kluwer.

Courbois, Y., Blades, M., Farran, E. K., & Sockeel, P. (2013). Do individuals with intellectual disability select appropriate objects as landmarks when learning a new route? *Journal of Intellectual Disability Research, 57*, 80–89.

Courbois, Y., Farran, E. K., Lemahieu, A., Blades, M., Mengue-Topio, H., & Sockeel, P. (2013). Wayfinding behaviour in Down syndrome: A study with virtual environments. *Research in Developmental Disabilities, 34*, 1825–1831.

Critchley, M. (1953). *The parietal lobes.* London: Edward Arnold.

Cursaru, G. (2017). *Parcourir l'invisible: Les espaces insondables à travers les déplacements des dieux dans la pensée religieuse grecque de l'époque archaïque* [Walking through the invisible: Unfathomable spaces through gods' travels in Greek religious thought of the Archaic period]. Louvain, Belgium: Peeters.

Cushman, L. A., Stein, K., & Duffy, C. J. (2008). Detecting navigational deficits in cognitive aging and Alzheimer disease using virtual reality. *Neurology, 71*, 888–895.

Danckert, J., & Ferber, S. (2006). Revisiting unilateral neglect. *Neuropsychologia, 44*, 987–1006.

Daniel, M.-P., & Denis, M. (2004). The production of route directions: Investigating conditions that favour conciseness in spatial discourse. *Applied Cognitive Psychology, 18*, 57–75.

Daniel, M.-P., Tom, A., Manghi, E., & Denis, M. (2003). Testing the value of route directions through navigational performance. *Spatial Cognition and Computation, 3*, 269–289.

Dasen, P. R., & Mishra, R. C. (2010). *Development of geocentric spatial language and cognition: An eco-cultural perspective*. Cambridge, UK: Cambridge University Press.

De Goede, M., & Postma, A. (2015). Learning your way in a city: Experience and gender differences in configurational knowledge of one's environment. *Frontiers in Psychology, 6*: 402.

Dehaene, S. (1997). *The number sense: How the mind creates mathematics*. New York: Oxford University Press.

Dehaene, S., Bossini, S., & Giraux, P. (1993). The mental representation of parity and number magnitude. *Journal of Experimental Psychology: General, 122*, 371–396.

Dehaene, S., Izard, V., Pica, P., & Spelke, E. (2006). Core knowledge of geometry in an Amazonian indigene group. *Science, 311*, 381–384.

de Hevia, M. D., Girelli, L., & Vallar, G. (2006). Numbers and space: A cognitive illusion? *Experimental Brain Research, 168*, 254–264.

de Hevia, M. D., & Spelke, E. S. (2009). Spontaneous mapping of number and space in adults and young children. *Cognition, 110*, 198–207.

deIpolyi, A. R., Rankin, K. P., Mucke, L., Miller, B. L., & Gorno-Tempini, M. L. (2007). Spatial cognition and the human navigation network in AD and MCI. *Neurology, 69*, 986–997.

Denis, M. (1996). Imagery and the description of spatial configurations. In M. de Vega, M. J. Intons-Peterson, P. N. Johnson-Laird, M. Denis, & M. Marschark, *Models of visuospatial cognition* (pp. 128–197). New York: Oxford University Press.

Denis, M. (1997). The description of routes: A cognitive approach to the production of spatial discourse. *Current Psychology of Cognition, 16*, 409–458.

Denis, M. (2008). Assessing the symbolic distance effect in mental images constructed from verbal descriptions: A study of individual differences in the mental comparison of distances. *Acta Psychologica, 127*, 197–210.

Denis, M. (2018). Arrows in diagrammatic and navigational spaces. In J. M. Zacks & H. A. Taylor (Eds.), *Representations in mind and world: Essays inspired by Barbara Tversky* (pp. 63–84). New York: Routledge.

Denis, M., Beschin, N., Logie, R. H., & Della Sala, S. (2002). Visual perception and verbal descriptions as sources for generating mental representations: Evidence from representational neglect. *Cognitive Neuropsychology, 19*, 97–112.

Denis, M., & Carfantan, M. (1985). People's knowledge about images. *Cognition, 20*, 49–60.

Denis, M., & Cocude, M. (1989). Scanning visual images generated from verbal descriptions. *European Journal of Cognitive Psychology, 1*, 293–307.

Denis, M., & Cocude, M. (1992). Structural properties of visual images constructed from poorly or well-structured verbal descriptions. *Memory and Cognition, 20*, 497–506.

Denis, M., & Denhière, G. (1990). Comprehension and recall of spatial descriptions. *European Bulletin of Cognitive Psychology, 10*, 115–143.

Denis, M., & Donikian, S. (2009). Navigation et cognition spatiale: De l'étude de l'humain réel à la simulation de l'humain virtuel [Navigation and spatial cognition: From the study of the real human to the simulation of the virtual human]. In S. Donikian (Ed.), *Le traité de la réalité virtuelle*. Vol. 5: *Les humains virtuels* (pp. 285–301). Paris: Presses des Mines.

Denis, M., & Fernandez, G. (2013). The processing of landmarks in route directions. In T. Tenbrink, J. Wiener, & C. Claramunt (Eds.), *Representing space in cognition: Interrelations of behaviour, language, and formal models* (pp. 42–55). Oxford: Oxford University Press.

Denis, M., Gonçalves, M.-R., & Memmi, D. (1995). Mental scanning of visual images generated from verbal descriptions: Towards a model of image accuracy. *Neuropsychologia, 33*, 1511–1530.

Denis, M., & Kosslyn, S. M. (1999). Scanning visual mental images: A window on the mind. *Current Psychology of Cognition, 18*, 409–465.

Denis, M., Logie, R. H., Cornoldi, C., de Vega, M., & Engelkamp, J. (Eds.) (2001). *Imagery, language, and visuo-spatial thinking*. Hove, UK: Psychology Press.

Denis, M., Michon, P.-E., & Tom, A. (2006). Assisting pedestrian wayfinding in urban settings: Why references to landmarks are crucial in direction-giving. In G. L. Allen (Ed.), *Applied spatial cognition: From research to cognitive technology* (pp. 25–51). Mahwah, NJ: Erlbaum.

Denis, M., Mores, C., Gras, D., Gyselinck, V., & Daniel, M.-P. (2014). Is memory for routes enhanced by an environment's richness in visual landmarks? *Spatial Cognition and Computation, 14*, 284–305.

Denis, M., Pazzaglia, F., Cornoldi, C., & Bertolo, L. (1999). Spatial discourse and navigation: An analysis of route directions in the city of Venice. *Applied Cognitive Psychology, 13*, 145–174.

Denis, M., & Zimmer, H. D. (1992). Analog properties of cognitive maps constructed from verbal descriptions. *Psychological Research, 54*, 286–298.

Detienne, M. (1966). Mythe et pensée dans la Grèce ancienne: Problèmes de psychologie historique [Myth and thought in ancient Greece: Issues of historical psychology]. *Archives de Sociologie des Religions, 21*, 125–134.

de Vega, M. (1994). Characters and their perspectives in narratives describing spatial environments. *Psychological Research, 56*, 116–126.

de Vega, M., & Rodrigo, M. J. (2001). Updating spatial layouts mediated by pointing and labelling under physical and imaginary rotation. *European Journal of Cognitive Psychology, 13*, 369–393.

Deyzac, E., Logie, R. H., & Denis, M. (2006). Visuospatial working memory and the processing of spatial descriptions. *British Journal of Psychology, 97*, 217–243.

Dijkstra, E. W. (1959). A note on two problems in connexion with graphs. *Numerische Mathematik, 1*, 269–271.

Duckham, M., & Kulik, L. (2003). "Simplest" paths: Automated route selection for navigation. In W. Kuhn, M. F. Worboys, & S. Timpf (Eds.), *Spatial information theory: Foundations of geographic information science (Proceedings of COSIT 2003)* (pp. 169–185). Berlin: Springer.

Duckham, M., Winter, S., & Robinson, M. (2010). Including landmarks in routing instructions. *Journal of Location Based Services, 4*, 28–52.

Dudchenko, P. A. (2010). *Why people get lost: The psychology and neuroscience of spatial cognition*. Oxford: Oxford University Press.

Duran, N. D., Dale, R., & Kreuz, R. J. (2011). Listeners invest in an assumed other's perspective despite cognitive cost. *Cognition, 121*, 22–40.

Eco, U. (1983). *The name of the rose*. New York: Harcourt.

Eco, U. (2014). *From the tree to the labyrinth: Historical studies on the sign and interpretation*. Cambridge, MA: Harvard University Press.

Ehrich, V., & Koster, C. (1983). Discourse organization and sentence form: The structure of room descriptions in Dutch. *Discourse Processes, 6*, 169–195.

Ehrlich, K., & Johnson-Laird, P. N. (1982). Spatial descriptions and referential continuity. *Journal of Verbal Learning and Verbal Behavior, 21*, 296–306.

Einstein, A. (1920). *Relativity: The special and general theory*. New York: Henry Holt and Company.

Ekstrom, A. D., Kahana, M. J., Caplan, J. B., Fields, T. A., Isham, E. A., Newman, E. L., & Fried, I. (2003). Cellular networks underlying human spatial navigation. *Nature, 425*, 184–187.

Elden, S., & Mendieta, E. (Eds.) (2011). *Reading Kant's geography*. Albany, NY: State University of New York Press.

Eliade, M. (1959). *The sacred and the profane: The nature of religion*. New York: Harcourt, Brace & World.

Elias, B. (2003). Extracting landmarks with data mining methods. In W. Kuhn, M. F. Worboys, & S. Timpf (Eds.), *Spatial information theory: Foundations of geographic information science (Proceedings of COSIT 2003)* (pp. 375–389). Berlin: Springer.

Epstein, R. A. (2008). Parahippocampal and retrosplenial contributions to human spatial navigation. *Trends in Cognitive Sciences, 12*, 388–396.

Epstein, R., & Kanwisher, N. (1998). A cortical representation of the local visual environment. *Nature, 392*, 598–601.

Farah, M. J. (1989). The neural basis of mental imagery. *Trends in Neurosciences, 12*, 395–399.

Farran, E. K., Courbois, Y., Van Herwegen, J., & Blades, M. (2012). How useful are landmarks when learning a route in a virtual environment? Evidence from typical development and Williams syndrome. *Journal of Experimental Child Psychology, 111*, 571–586.

Farrell, M. J., Arnold, P., Pettifer, S., Adams, J., Graham, T., & MacManamon, M. (2003). Transfer of route learning from virtual to real environments. *Journal of Experimental Psychology: Applied, 9*, 219–227.

Fischer, M. H. (2001). Number processing induces spatial performance biases. *Neurology, 57*, 822–826.

Foos, P. W. (1980). Constructing cognitive maps from sentences. *Journal of Experimental Psychology: Human Learning and Memory, 6*, 25–38.

Frank, A. U. (2003). Pragmatic information content – How to measure the information in a route description. In M. Duckham, M. F. Goodchild, & M. F. Worboys (Eds.), *Foundations of geographic information science* (pp. 47–68). London: Taylor & Francis.

Frankenstein, J., Mohler, B. J., Bülthoff, H. H., & Meilinger, T. (2012). Is the map in our head oriented north? *Psychological Science, 23*, 120–125.

Franklin, N., & Tversky, B. (1990). Searching imagined environments. *Journal of Experimental Psychology: General, 119*, 63–76.

Franklin, N., Tversky, B., & Coon, V. (1992). Switching points of view in spatial mental models. *Memory and Cognition, 20*, 507–518.

Franz, M. O., & Mallot, H. A. (2000). Biomimetic robot navigation. *Robotics and Autonomous Systems, 30*, 133–153.

Franz, M. O., Stürzl, W., Hübner, W., & Mallot, H. A. (2008). A robot system for biomimetic navigation – From snapshots to metric embeddings of view graphs. In M. E. Jefferies & W.-K. Yeap (Eds.), *Robotics and cognitive approaches to spatial mapping* (pp. 297–314). Berlin: Springer.

Freksa, C. (1991). Qualitative spatial reasoning. In D. M. Mark & A. U. Frank (Eds.), *Cognitive and linguistic aspects of geographic space* (pp. 361–372). Dordrecht, The Netherlands: Kluwer.

Freksa, C. (2004). Spatial cognition: An AI perspective. In R. López de Mántaras & L. Saitta (Eds.), *Proceedings of the 16th European Conference on Artificial Intelligence (ECAI 2004)* (pp. 1122–1128). Amsterdam: IOS Press.

Freksa, C. (2013). Spatial computing: How spatial structures replace computational effort. In M. Raubal, D. M. Mark, & A. U. Frank (Eds.), *Cognitive and linguistic aspects of geographic space: New perspectives on geographic information research* (pp. 23–42). Berlin: Springer.

Freksa, C., Moratz, R., & Barkowsky, T. (2000). Schematic maps for robot navigation. In C. Freksa, C. Habel, W. Brauer, & K. F. Wender (Eds.), *Spatial cognition II: Integrating abstract theories, empirical studies, formal methods, and practical applications* (pp. 100–114). Berlin: Springer.

Frémont, A. (2005). Géographie et espace vécu [Geography and experienced space]. In A. Berthoz & R. Recht (Eds.), *Les espaces de l'homme* (pp. 93–107). Paris: Odile Jacob.

Frenz, H., Lappe, M., Kolesnik, M., & Bührmann, T. (2007). Estimation of travel distance from visual motion in virtual environments. *ACM Transactions on Applied Perception, 4*, Article 3.

Fujita, N., Klatzky, R. L., Loomis, J. M., & Golledge, R. G. (1993). The encoding-error model of pathway completion without vision. *Geographical Analysis, 25*, 295–314.

Galea, L. A. M., & Kimura, D. (1993). Sex differences in route-learning. *Personality and Individual Differences, 14*, 53–65.

Gallay, M., Denis, M., & Auvray, M. (2013). Navigation assistance for blind pedestrians: Guidelines for the design of devices and implications for spatial cognition. In T. Tenbrink, J. Wiener, & C. Claramunt (Eds.), *Representing space in cognition: Interrelations of behaviour, language, and formal models* (pp. 244–267). Oxford: Oxford University Press.

Galton, F. (1880). Visualised numerals. *Nature, 21*, 252–256, 494–495.

Garden, S., Cornoldi, C., & Logie, R. H. (2002). Visuo-spatial working memory in navigation. *Applied Cognitive Psychology, 16*, 35–50.

Gardony, A. L., Brunyé, T. T., Mahoney, C. R., & Taylor, H. A. (2013). How navigational aids impair spatial memory: Evidence for divided attention. *Spatial Cognition and Computation, 13*, 319–350.

Gärling, T., & Golledge, R. G. (1989). Environmental perception and cognition. In E. H. Zube & G. T. Moore (Eds.), *Advances in environment, behavior, and design* (Vol. 2, pp. 203–236). New York: Plenum.

Gaussier, P., & Zrehen, S. (1995). PerAc: A neural architecture to control artificial animals. *Robotics and Autonomous Systems, 16*, 291–320.

Gell, A. (1985). How to read a map: Remarks on the practical logic of navigation. *Man, 20*, 271–286.

Getis, A., & Paelinck, J. (2004). An analytical description of spatial patterns. *L'Espace Géographique, 33*, 61–68.

Gevers, W., Reynvoet, B., & Fias, W. (2003). The mental representation of ordinal sequences is spatially organized. *Cognition, 87*, B87–B95.

Ghaëm, O., Mellet, E., Crivello, F., Tzourio, N., Mazoyer, B., Berthoz, A., & Denis, M. (1997). Mental navigation along memorized routes activates the hippocampus, precuneus, and insula. *NeuroReport, 8*, 739–744.

Gibson, J. J. (1950). *The perception of the visual world.* Cambridge, MA: The Riverside Press.

Gillner, S., & Mallot, H. A. (1998). Navigation and acquisition of spatial knowledge in a virtual maze. *Journal of Cognitive Neuroscience, 10*, 445–463.

Giraudo, M.-D., & Pailhous, J. (1994). Distortions and fluctuations in topographic memory. *Memory and Cognition, 22*, 14–26.

Giraut, F. (2013). Territoire multisitué, complexité territoriale et postmodernité territoriale: Des concepts opératoires pour rendre compte des territorialités contemporaines? [Multilocated territory, territorial complexity and territorial postmodernity: Operational concepts to account for today's territorialities?] *L'Espace Géographique, 42*, 293–305.

Glass, A. L. (2016). *Cognition: A neuroscience approach.* Cambridge, UK: Cambridge University Press.

Golding, J. M., Graesser, A. C., & Hauselt, J. (1996). The process of answering direction-giving questions when someone is lost on a university campus: The role of pragmatics. *Applied Cognitive Psychology, 10*, 23–39.

Golledge, R. G. (1992). Place recognition and wayfinding: Making sense of space. *Geoforum, 23*, 199–214.

Golledge, R. G. (1999). Human wayfinding and cognitive maps. In R. G. Golledge (Ed.), *Wayfinding behavior: Cognitive mapping and other spatial processes* (pp. 5–45). Baltimore, MD: Johns Hopkins University Press.

Götze, J., & Boye, J. (2016). Learning landmark salience models from users' route instructions. *Journal of Location Based Services, 10*, 47–63.

Gracq, J. (1986). *The opposing shore*. New York: Columbia University Press.

Gray Wilson, S., Rinck, M., McNamara, T. P., Bower, G. H., & Morrow, D. G. (1993). Mental models and narrative comprehension: Some qualifications. *Journal of Memory and Language, 32*, 141–154.

Grésillon, A., Lebrave, J.-L., & Fuchs, C. (1991). Flaubert: "Ruminer Hérodias." Du cognitif-visuel au verbal-textuel [Flaubert: "Ruminating Hérodias." From visual-cognitive to textual-verbal]. In D. Ferrer & J.-L. Lebrave (Eds.), *L'écriture et ses doubles: Genèse et variation textuelle* (pp. 27–109). Paris: Editions du CNRS.

Grice, H. P. (1975). Logic and conversation. In P. Cole & J. L. Morgan (Eds.), *Syntax and semantics*. Vol. 3: *Speech acts* (pp. 41–58). New York: Academic Press.

Guariglia, C., Padovani, A., Pantano, P., & Pizzamiglio, L. (1993). Unilateral neglect restricted to visual imagery. *Nature, 364*, 235–237.

Guérin-Pace, F. (2006). Sentiment d'appartenance et territoires identitaires [Feeling of membership and identity territories]. *L'Espace Géographique, 35*, 298–308.

Gutruf, G., & Stachel, H. (2010). The hidden geometry in Vermeer's "The Art of Painting." *Journal of Geometry and Graphics, 14*, 187–202.

Gyselinck, V., Meneghetti, C., De Beni, R., & Pazzaglia, F. (2009). The role of working memory in spatial text processing: What benefit of imagery strategy and visuospatial abilities? *Learning and Individual Differences, 19*, 12–20.

Gyselinck, V., & Pazzaglia, F. (Eds.) (2012). *From mental imagery to spatial cognition and language: Essays in honour of Michel Denis*. Hove, UK: Psychology Press.

Haber, R. N., Haber, L. R., Levin, C. A., & Hollyfield, R. (1993). Properties of spatial representations: Data from sighted and blind subjects. *Perception and Psychophysics, 54*, 1–13.

Hafting, T., Fyhn, M., Molden, S., Moser, M.-B., & Moser, E. I. (2005). Microstructure of a spatial map in the entorhinal cortex. *Nature, 436*, 801–806.

Hao, X., Huang, Y., Li, X., Song, Y., Kong, X., Wang, X., Yang, Z., Zhen, Z., & Liu, J. (2016). Structural and functional neural correlates of spatial navigation: A combined voxel-based morphometry and functional connectivity study. *Brain and Behavior, 6*, e00572.

Haun, D. B. M., Rapold, C. J., Janzen, G., & Levinson, S. C. (2011). Plasticity of human spatial cognition: Spatial language and cognition covary across cultures. *Cognition, 119*, 70–80.

Haviland, J. B. (1998). Guugu Yimithirr cardinal directions. *Ethos, 26*, 25–47.

He, G., Ishikawa, T., & Takemiya, M. (2015). Collaborative navigation in an unfamiliar environment with people having different spatial aptitudes. *Spatial Cognition and Computation, 15*, 285–307.

Heft, H. (1979). The role of environmental features in route-learning: Two exploratory studies of way-finding. *Environmental Psychology and Nonverbal Behavior, 3*, 172–185.

Hegarty, M., Montello, D. R., Richardson, A. E., Ishikawa, T., & Lovelace, K. (2006). Spatial abilities at different scales: Individual differences in aptitude-test performance and spatial-layout learning. *Intelligence, 34*, 151–176.

Hegarty, M., Richardson, A. E., Montello, D. R., Lovelace, K., & Subbiah, I. (2002). Development of a self-report measure of environmental spatial ability. *Intelligence, 30*, 425–447.

Hegarty, M., & Waller, D. (2005). Individual differences in spatial abilities. In P. Shah & A. Miyake (Eds.), *The Cambridge handbook of visuospatial thinking* (pp. 121-169). Cambridge, MA: Cambridge University Press.

Herskovits, A. (1986). *Language and spatial cognition: An interdisciplinary study of the prepositions in English.* Cambridge, UK: Cambridge University Press.

Heth, C. D., Cornell, E. H., & Flood, T. L. (2002). Self-ratings of sense of direction and route reversal performance. *Applied Cognitive Psychology, 16*, 309–324.

Heuten, W., Henze, N., Boll, S., & Pielot, M. (2008). Tactile Wayfinder: A non-visual support system for wayfinding. In K. Tollmar & B. Jönsson (Eds.), *Proceedings of the 5th Nordic Conference on Human-Computer Interaction* (pp. 172–181). New York: ACM.

Hickmann, M., & Robert, S. (Eds.) (2006). *Space in languages: Linguistic systems and cognitive categories.* Amsterdam: John Benjamins.

Hill, C. (1982). Up/down, front/back, left/right: A contrastive study of Hausa and English. In J. Weissenborn & W. Klein (Eds.), *Here and there: Cross-linguistic studies on deixis and demonstration* (pp. 13–42). Amsterdam: John Benjamins.

Hill, E. W., Rieser, J. J., Hill, M.-M., Hill, M., Halpin, J., & Halpin, R. (1993). How persons with visual impairments explore novel spaces: Strategies of good and poor performers. *Journal of Visual Impairment and Blindness, 87*, 295–301.

Hirtle, S. C. (2008). Landmarks for navigation in human and robots. In M. E. Jefferies & W.-K. Yeap (Eds.), *Robotics and cognitive approaches to spatial mapping* (pp. 203–214). Berlin: Springer.

Hirtle, S. C., & Raubal, M. (2013). Many to many mobile maps. In M. Raubal, D. M. Mark, & A. U. Frank (Eds.), *Cognitive and linguistic aspects of geographic space: New perspectives on geographic information research* (pp. 141–157). Berlin: Springer.

Hirtle, S., Richter, K.-F., Srinivas, S., & Firth, R. (2010). This is the tricky part: When directions become difficult. *Journal of Spatial Information Science*, No. 1, 53–73.

Hollyfield, R. L., & Foulke, E. (1983). The spatial cognition of blind pedestrians. *Journal of Visual Impairment and Blindness*, 77, 204–210.

Hölscher, C., Schnee, A., Dahmen, H., Setia, L., & Mallot, H. A. (2005). Rats are able to navigate in virtual environments. *Journal of Experimental Biology, 208*, 561–569.

Hölscher, C., Tenbrink, T., & Wiener, J. M. (2011). Would you follow your own route description? Cognitive strategies in urban route planning. *Cognition, 121*, 228–247.

Horner, A. J., Bisby, J. A., Zotow, E., Bush, D., & Burgess, N. (2016). Grid-like processing of imagined navigation. *Current Biology, 26*, 842–847.

Hubbard, E. M., Piazza, M., Pinel, P., & Dehaene, S. (2005). Interactions between number and space in parietal cortex. *Nature Reviews Neuroscience, 6*, 435–448.

Hund, A. M., & Minarik, J. L. (2006). Getting from here to there: Spatial anxiety, wayfinding strategies, direction type, and wayfinding efficiency. *Spatial Cognition and Computation, 6*, 179–201.

Hund, A. M., & Padgitt, A. J. (2010). Direction giving and following in the service of wayfinding in a complex indoor environment. *Journal of Environmental Psychology, 30*, 553–564.

Iachini, T., Ruggiero, G., Conson, M., & Trojano, L. (2009). Lateralization of egocentric and allocentric spatial processing after parietal brain lesions. *Brain and Cognition, 69*, 514–520.

Iachini, T., Ruggiero, G., & Ruotolo, F. (2014). Does blindness affect egocentric and allocentric frames of reference in small and large scale spaces? *Behavioural Brain Research, 273*, 73–81.

Iaria, G., & Barton, J. J. S. (2010). Developmental topographical disorientation: A newly discovered cognitive disorder. *Experimental Brain Research, 206*, 189–196.

Iaria, G., Bogod, N., Fox, C. J., & Barton, J. J. S. (2009). Developmental topographical disorientation: Case one. *Neuropsychologia, 47*, 30–40.

Iaria, G., Chen, J.-K., Guariglia, C., Ptito, A., & Petrides, M. (2007). Retrosplenial and hippocampal brain regions in human navigation: Complementary functional contributions to the formation and use of cognitive maps. *European Journal of Neuroscience, 25*, 890–899.

Iaria, G., Petrides, M., Dagher, A., Pike, B., & Bohbot, V. D. (2003). Cognitive strategies dependent on the hippocampus and caudate nucleus in human navigation: Variability and change in practice. *Journal of Neuroscience, 23*, 5945–5952.

Iglói, K., Doeller, C. F., Berthoz, A., Rondi-Reig, L., & Burgess, N. (2010). Lateralized human hippocampal activity predicts navigation based on sequence or place memory. *Proceedings of the National Academy of Sciences, 107*, 14466–14471.

Iglói, K., Zaoui, M., Berthoz, A., & Rondi-Reig, L. (2009). Sequential egocentric strategy is acquired as early as allocentric strategy: Parallel acquisition of these two navigation strategies. *Hippocampus, 19*, 1199–1211.

Ishikawa, T., Fujiwara, H., Imai, O., & Okabe, A. (2008). Wayfinding with a GPS-based mobile navigation system: A comparison with maps and direct experience. *Journal of Environmental Psychology, 28*, 74–82.

Ishikawa, T., & Montello, D. R. (2006). Spatial knowledge acquisition from direct experience in the environment: Individual differences in the development of metric knowledge and the integration of separately learned places. *Cognitive Psychology, 52*, 93–129.

Ishikawa, T., & Nakamura, U. (2012). Landmark selection in the environment: Relationships with object characteristics and sense of direction. *Spatial Cognition and Computation, 12*, 1–22.

Izard, V., Pica, P., Dehaene, S., Hinchey, D., & Spelke, E. (2011). Geometry as a universal mental construction. In S. Dehaene & E. Brannon (Eds.), *Space, time and number in the brain: Searching for the foundations of mathematical thought* (pp. 319–332). London: Academic Press.

Jansen, P., Schmelter, A., & Heil, M. (2010). Spatial knowledge acquisition in younger and elderly adults: A study in a virtual environment. *Experimental Psychology, 57*, 54–60.

Jansen-Osmann, P. (2002). Using desktop virtual environments to investigate the role of landmarks. *Computers in Human Behavior, 18*, 427–436.

Jansen-Osmann, P., & Berendt, B. (2005). What makes a route appear longer? An experimental perspective on features, route segmentation, and distance knowledge. *Quarterly Journal of Experimental Psychology, 58A*, 1390–1414.

Jansen-Osmann, P., & Wiedenbauer, G. (2004). The representation of landmarks and routes in children and adults: A study in a virtual environment. *Journal of Environmental Psychology, 24*, 347–357.

Janzen, G., & van Turennout, M. (2004). Selective neural representation of objects relevant for navigation. *Nature Neuroscience, 7*, 673–677.

Janzen, G., Wagensveld, B., & van Turennout, M. (2007). Neural representation of navigational relevance is rapidly induced and long lasting. *Cerebral Cortex, 17*, 975–981.

Jeannerod, M. (2011). *La fabrique des idées* [The making of ideas]. Paris: Odile Jacob.

Johnson-Laird, P. N. (1983). *Mental models: Towards a cognitive science of language, inference, and consciousness*. Cambridge, UK: Cambridge University Press.

Kalakoski, V., & Saariluoma, P. (2001). Taxi drivers' exceptional memory of street names. *Memory and Cognition, 29*, 634–638.

Karadimas, D. (2005). Chercher le centre: Stratégie d'orientation spatiale chez les Miraña d'Amazonie colombienne [Looking for center: Spatial orientation strategies in

Miraña from Colombian Amazon]. In A. Berthoz & R. Recht (Eds.), *Les espaces de l'homme* (pp. 67–91). Paris: Odile Jacob.

Karimpur, H., Röser, F., & Hamburger, K. (2016). Finding the return path: Landmark position effects and the influence of perspective. *Frontiers in Psychology*, 7: 1956.

Kato, Y., & Takeuchi, Y. (2003). Individual differences in wayfinding strategies. *Journal of Environmental Psychology, 23*, 171–188.

Kavanaugh, L. (2008). Brentano on space. *Architecture and Phenomenology*, No. 3, 39–50.

Kesner, R. P., & Creem-Regehr, S. H. (2013). Parietal contributions to spatial cognition. In D. Waller & L. Nadel (Eds.), *Handbook of spatial cognition* (pp. 35–63). Washington, DC: American Psychological Association.

Klatzky, R. L. (1998). Allocentric and egocentric spatial representations: Definitions, distinctions, and interconnections. In C. Freksa, C. Habel, & K. F. Wender (Eds.), *Spatial cognition: An interdisciplinary approach to representing and processing spatial knowledge* (pp. 1–17). Berlin: Springer.

Klatzky, R. L., Golledge, R. G., Loomis, J. M., Cicinelli, J. G., & Pellegrino, J. W. (1995). Performance of blind and sighted persons on spatial tasks. *Journal of Visual Impairment and Blindness, 89*, 70–82.

Klatzky, R. L., Lippa, Y., Loomis, J. M., & Golledge, R. G. (2003). Encoding, learning, and spatial updating of multiple object locations specified by 3-D sound, spatial language, and vision. *Experimental Brain Research, 149*, 48–61.

Klatzky, R. L., Loomis, J. M., & Golledge, R. G. (1997). Encoding spatial representations through nonvisually guided locomotion: Tests of human path integration. In D. Medin (Ed.), *The psychology of learning and motivation* (Vol. 37, pp. 41–84). San Diego, CA: Academic Press.

Klatzky, R. L., Loomis, J. M., Golledge, R. G., Cicinelli, J. G., Doherty, S., & Pellegrino, J. W. (1990). Acquisition of route and survey knowledge in the absence of vision. *Journal of Motor Behavior, 22*, 19–43.

Klatzky, R. L., Marston, J. R., Giudice, N. A., Golledge, R. G., & Loomis, J. M. (2006). Cognitive load of navigating without vision when guided by virtual sound versus spatial language. *Journal of Experimental Psychology: Applied, 12*, 223–232.

Klein, W. (1982). Local deixis in route directions. In R. J. Jarvella & W. Klein (Eds.), *Speech, place, and action: Studies in deixis and related topics* (pp. 161–182). Chichester, UK: Wiley.

Klein, W. (1993). Some notorious pitfalls in the analysis of spatial expressions. In F. Beckmann & G. Heyer (Eds.), *Theorie und Praxis des Lexikons* (pp. 191–204). Berlin: de Gruyter.

Klippel, A. (2003). Wayfinding choremes. In W. Kuhn, M. F. Worboys, & S. Timpf (Eds.), *Spatial information theory: Foundations of geographic information science (Proceedings of COSIT 2003)* (pp. 301–315). Berlin: Springer.

Klippel, A., Tappe, H., Kulik, L., & Lee, P. U. (2005). Wayfinding choremes – A language for modeling conceptual route knowledge. *Journal of Visual Languages and Computing, 16*, 311–329.

Klippel, A., & Winter, S. (2005). Structural salience of landmarks for route directions. In A. G. Cohn & D. M. Mark (Eds.), *Spatial information theory (Proceedings of COSIT 2005)* (pp. 347–362). Berlin: Springer.

Knauff, M. (2013). *Space to reason: A spatial theory of human thought*. Cambridge, MA: The MIT Press.

Kosslyn, S. M., Ball, T. M., & Reiser, B. J. (1978). Visual images preserve metric spatial information: Evidence from studies of image scanning. *Journal of Experimental Psychology: Human Perception and Performance, 4*, 47–60.

Kosslyn, S. M., Reiser, B. J., Farah, M. J., & Fliegel, S. L. (1983). Generating visual images: Units and relations. *Journal of Experimental Psychology: General, 112*, 278–303.

Kozhevnikov, M., & Hegarty, M. (2001). A dissociation between object manipulation spatial ability and spatial orientation ability. *Memory and Cognition, 29*, 745–756.

Kozhevnikov, M., Motes, M. A., Rasch, B., & Blajenkova, O. (2006). Perspective-taking vs. mental rotation transformations and how they predict spatial navigation performance. *Applied Cognitive Psychology, 20*, 397–417.

Kuipers, B. (1978). Modeling spatial knowledge. *Cognitive Science, 2*, 129–153.

Kuipers, B. (2000). The Spatial Semantic Hierarchy. *Artificial Intelligence, 119*, 191–233.

Kuipers, B. J., & Byun, Y.-T. (1991). A robot exploration and mapping strategy based on a semantic hierarchy of spatial representations. *Robotics and Autonomous Systems, 8*, 47–63.

Kuipers, B., Tecuci, D. G., & Stankiewicz, B. J. (2003). The skeleton in the cognitive map: A computational and empirical exploration. *Environment and Behavior, 35*, 81–106.

Lacoste, Y. (1987). Julien Gracq, un écrivain géographe: *Le Rivage des Syrtes*, un roman géopolitique [Julien Gracq, writer and geographer: *The Opposing Shore*, a geopolitical novel]. *Hérodote*, No. 44, 8–37.

Landau, B., & Jackendoff, R. (1993). "What" and "where" in spatial language and spatial cognition. *Behavioral and Brain Sciences, 16*, 217–265.

Latini Corazzini, L., Nesa, M. P., Ceccaldi, M., Guedj, E., Thinus-Blanc, C., Cauda, F., Dagata, F., & Péruch, P. (2010). Route and survey processing of topographical memory during navigation. *Psychological Research, 74*, 545–559.

Lawton, C. A. (1994). Gender differences in way-finding strategies: Relationship to spatial ability and spatial anxiety. *Sex Roles, 30*, 765–779.

Lawton, C. A. (1996). Strategies for indoor wayfinding: The role of orientation. *Journal of Environmental Psychology, 16*, 137–145.

Lawton, C. A., & Kallai, J. (2002). Gender differences in wayfinding strategies and anxiety about wayfinding: A cross-cultural comparison. *Sex Roles, 47*, 389–401.

Leighty, K. A., & Fragaszy, D. M. (2003). Primates in cyberspace: Using interactive computer tasks to study perception and action in nonhuman animals. *Animal Cognition, 6*, 137–139.

Levelt, W. J. M. (1982). Linearization in describing spatial networks. In S. Peters & E. Saarinen (Eds.), *Processes, beliefs, and questions* (pp. 199–220). Dordrecht, The Netherlands: Reidel.

Levelt, W. J. M. (1989). *Speaking: From intention to articulation.* Cambridge, MA: The MIT Press.

Levelt, W. J. M. (1996). Perspective taking and ellipsis in spatial descriptions. In P. Bloom, M. A. Peterson, L. Nadel, & M. F. Garrett (Eds.), *Language and space* (pp. 77–107). Cambridge, MA: The MIT Press.

Levinson, S. C. (1996). Frames of reference and Molyneux's question: Crosslinguistic evidence. In P. Bloom, M. A. Peterson, L. Nadel, & M. F. Garrett (Eds.), *Language and space* (pp. 109–169). Cambridge, MA: The MIT Press.

Levinson, S. C. (2003). *Space in language and cognition: Explorations in cognitive diversity.* Cambridge, UK: Cambridge University Press.

Lévy, J. (2004). La cartographie, enjeu contemporain [Cartography: A challenge of today]. *Documentation Photographique*, No. 8036, 1–16.

Lewis, D. (1976). Observations on route finding and spatial orientation among the Aboriginal peoples of the Western Desert region of Central Australia. *Oceania, 46*, 249–282.

Li, R., & Klippel, A. (2012). Wayfinding in libraries: Can problems be predicted? *Journal of Map and Geography Libraries, 8*, 21–38.

Ligozat, G. F. (1993). Qualitative triangulation for spatial reasoning. In A. U. Frank & I. Campari (Eds.), *Spatial information theory: A theoretical basis for GIS (Proceedings of COSIT 1993)* (pp. 54–68). Berlin: Springer.

Ligozat, G. (2012). *Qualitative spatial and temporal reasoning*. London: ISTE / Hoboken, NJ: Wiley.

Linde, C., & Labov, W. (1975). Spatial networks as a site for the study of language and thought. *Language, 51*, 924–939.

Lithfous, S., Dufour, A., & Després, O. (2013). Spatial navigation in normal aging and the prodromal stage of Alzheimer's disease: Insights from imaging and behavioral studies. *Ageing Research Reviews, 12*, 201–213.

Liu, I., Levy, R. M., Barton, J. J. S., & Iaria, G. (2011). Age and gender differences in various topographical orientation strategies. *Brain Research, 1410*, 112–119.

Lloyd, P. (1991). Strategies used to communicate route directions by telephone: A comparison of the performance of 7-year-olds, 10-year-olds and adults. *Journal of Child Language, 18*, 171–189.

Logie, R. H. (1995). *Visuo-spatial working memory*. Hove, UK: Erlbaum.

Longo, G. (2003). Géométrie et cognition: Entre fondements des mathématiques, théorie de la connaissance et cognition [Geometry and cognition: Between foundations of mathematics, theory of knowledge and cognition]. *Revue de Synthèse, 124*, 1–10.

Loomis, J. M., Da Silva, J. A., Fujita, N., & Fukusima, S. S. (1992). Visual space perception and visually directed action. *Journal of Experimental Psychology: Human Perception and Performance, 18*, 906–921.

Loomis, J. M., Klatzky, R. L., & Giudice, N. A. (2013). Representing 3D space in working memory: Spatial images from vision, hearing, touch, and language. In S. Lacey & R. Lawson (Eds.), *Multisensory imagery* (pp. 131–155). New York: Springer.

Loomis, J. M., Klatzky, R. L., Golledge, R. G., Cicinelli, J. G., Pellegrino, J. W., & Fry, P. A. (1993). Nonvisual navigation by blind and sighted: Assessment of path integration ability. *Journal of Experimental Psychology: General, 122*, 73–91.

Loomis, J. M., Klatzky, R. L., Golledge, R. G., & Philbeck, J. W. (1999). Human navigation by path integration. In R. G. Golledge (Ed.), *Wayfinding behavior: Cognitive mapping and other spatial processes* (pp. 125–151). Baltimore, MD: Johns Hopkins University Press.

Loomis, J. M., Lippa, Y., Klatzky, R. L., & Golledge, R. G. (2002). Spatial updating of locations specified by 3-D sound and spatial language. *Journal of Experimental Psychology: Learning, Memory, and Cognition, 28*, 335–345.

Loomis, J. M., Marston, J. R., Golledge, R. G., & Klatzky, R. L. (2005). Personal guidance system for people with visual impairment: A comparison of spatial displays for route guidance. *Journal of Visual Impairment and Blindness, 99*, 219–232.

Lovecraft, H. P. (1976). *To Quebec and the stars*. Hampton Falls, NH: Donald M. Grant.

Lovito, G. (2013). Le mythe du labyrinthe revisité par Eco théoricien et romancier à des fins cognitives et métaphoriques [The myth of the labyrinth revisited by Eco theoretician and writer for cognitive and metaphorical purposes]. *Cahiers d'Etudes Romanes*, No. 27, 345–357.

Lussault, M. (2007). *L'homme spatial: La construction sociale de l'espace humain* [Spatial man: The social construction of human space]. Paris: Editions du Seuil.

Lynch, K. (1960). *The image of the city*. Cambridge, MA: The MIT Press.

Madsen, C. B., & Andersen, C. S. (1998). Optimal landmark selection for triangulation of robot position. *Robotics and Autonomous Systems, 23*, 277–292.

Maguire, E. A., Frackowiak, R. S. J., & Frith, C. D. (1997). Recalling routes around London: Activation of the right hippocampus in taxi drivers. *Journal of Neuroscience, 17*, 7103–7110.

Maguire, E. A., Gadian, D. G., Johnsrude, I. S., Good, C. D., Ashburner, J., Frackowiak, R. S. J., & Frith, C. D. (2000). Navigation-related structural change in the hippocampi of taxi drivers. *Proceedings of the National Academy of Sciences, 97*, 4398–4403.

Maguire, E. A., Nannery, R., & Spiers, H. J. (2006). Navigation around London by a taxi driver with bilateral hippocampal lesions. *Brain, 129*, 2894–2907.

Maguire, E. A., Spiers, H. J., Good, C. D., Hartley, T., Frackowiak, R. S. J., & Burgess, N. (2003). Navigation expertise and the human hippocampus: A structural brain imaging analysis. *Hippocampus, 13*, 250–259.

Majid, A., Bowerman, M., Kita, S., Haun, D. B. M., & Levinson, S. C. (2004). Can language restructure cognition? The case for space. *Trends in Cognitive Sciences, 8*, 108–114.

Maki, R. H. (1981). Categorization and distance effects with spatial linear orders. *Journal of Experimental Psychology: Human Learning and Memory, 7*, 15–32.

Mallot, H. A. (1997). Behavior-oriented approaches to cognition: Theoretical perspectives. *Theory in Biosciences, 116*, 196–220.

Mallot, H. A., & Gillner, S. (2000). Route navigating without place recognition: What is recognised in recognition-triggered responses? *Perception, 29*, 43–55.

Mani, K., & Johnson-Laird, P. N. (1982). The mental representation of spatial descriptions. *Memory and Cognition, 10*, 181–187.

Marchette, S. A., Vass, L. K., Ryan, J., & Epstein, R. A. (2014). Anchoring the neural compass: Coding of local spatial reference frames in human medial parietal lobe. *Nature Neuroscience, 17*, 1598–1606.

Mark, D. M. (1985). Finding simple routes: "Ease of description" as an objective function in automated route selection. In *Proceedings of the Second Symposium on Artificial Intelligence Applications* (pp. 577–581). Los Alamitos, CA: IEEE Computer Society Press.

Marston, J. R., Loomis, J. M., Klatzky, R. L., & Golledge, R. G. (2007). Nonvisual route following with guidance from a simple haptic or auditory display. *Journal of Visual Impairment and Blindness, 101*, 203–211.

Marston, J. R., Loomis, J. M., Klatzky, R. L., Golledge, R. G., & Smith, E. L. (2006). Evaluation of spatial displays for navigation without sight. *ACM Transactions on Applied Perception, 3*, 110–124.

May, M., & Klatzky, R. L. (2000). Path integration while ignoring irrelevant movement. *Journal of Experimental Psychology: Learning, Memory, and Cognition, 26*, 169–186.

McNamara, T. P. (2003). How are the locations of objects in the environment represented in memory? In C. Freksa, W. Brauer, C. Habel, & K. F. Wender (Eds.), *Spatial cognition III: Routes and navigation, human memory and learning, spatial representation and spatial reasoning* (pp. 174–191). Berlin: Springer.

McNamara, T. P. (2013). Spatial memory: Properties and organization. In D. Waller & L. Nadel (Eds.), *Handbook of spatial cognition* (pp. 173–190). Washington, DC: American Psychological Association.

Meilinger, T., Knauff, M., & Bülthoff, H. H. (2008). Working memory in wayfinding – A dual task experiment in a virtual city. *Cognitive Science, 32*, 755–770.

Meilinger, T., Schulte-Pelkum, J., Frankenstein, J., Hardiess, G., Laharnar, N., Mallot, H. A., & Bülthoff, H. H. (2016). How to best name a place? Facilitation and inhibition of route learning due to descriptive and arbitrary location labels. *Frontiers in Psychology, 7*: 76.

Mellet, E., Bricogne, S., Crivello, F., Mazoyer, B., Denis, M., & Tzourio-Mazoyer, N. (2002). Neural basis of mental scanning of a topographic representation built from a text. *Cerebral Cortex, 12*, 1322–1330.

Mellet, E., Bricogne, S., Tzourio-Mazoyer, N., Ghaëm, O., Petit, L., Zago, L., Etard, O., Berthoz, A., Mazoyer, B., & Denis, M. (2000). Neural correlates of topographic mental exploration: The impact of route versus survey perspective learning. *NeuroImage, 12*, 588–600.

Mellet, E., Laou, L., Petit, L., Zago, L., Mazoyer, B., & Tzourio-Mazoyer, N. (2010). Impact of the virtual reality on the neural representation of an environment. *Human Brain Mapping, 31*, 1065–1075.

Mellet, E., Tzourio-Mazoyer, N., Bricogne, S., Mazoyer, B., Kosslyn, S. M., & Denis, M. (2000). Functional anatomy of high-resolution visual mental imagery. *Journal of Cognitive Neuroscience, 12*, 98–109.

Meneghetti, C., Pazzaglia, F., & De Beni, R. (2011). Spatial mental representations derived from survey and route descriptions: When individuals prefer extrinsic frame of reference. *Learning and Individual Differences, 21*, 150–157.

Mengue-Topio, H., Courbois, Y., Farran, E. K., & Sockeel, P. (2011). Route learning and shortcut performance in adults with intellectual disability: A study with virtual environments. *Research in Developmental Disabilities, 32*, 345–352.

Merleau-Ponty, M. (1962). *Phenomenology of perception.* London: Routledge & Kegan Paul.

Merleau-Ponty, M. (1993). Eye and mind. In G. A. Johnson (Ed.), *The Merleau-Ponty aesthetics reader: Philosophy and painting* (pp. 121–149). Evanston, IL: Northwestern University Press.

Michon, P.-E., & Denis, M. (2001). When and why are visual landmarks used in giving directions? In D. R. Montello (Ed.), *Spatial information theory: Foundations of geographic information science (Proceedings of COSIT 2001)* (pp. 292–305). Berlin: Springer.

Miller, A. I. (1996). *Insights of genius: Imagery and creativity in science and art.* New York: Copernicus.

Moar, I., & Bower, G. H. (1983). Inconsistency in spatial knowledge. *Memory and Cognition, 11*, 107–113.

Moar, I., & Carleton, L. R. (1982). Memory for routes. *Quarterly Journal of Experimental Psychology, 34A*, 381–394.

Moffat, S. D. (2009). Aging and spatial navigation: What do we know and where do we go? *Neuropsychology Review, 19*, 478–489.

Moles, A., & Rohmer, E. (1998). *Psychosociologie de l'espace* [Psychosociology of space]. Paris: L'Harmattan.

Mollerup, P. (2013). *Wayshowing>wayfinding: Basic and interactive.* Amsterdam: BIS.

Monmonier, M. (1991). *How to lie with maps.* Chicago, IL: University of Chicago Press.

Montello, D. R. (1993). Scale and multiple psychologies of space. In A. U. Frank & I. Campari (Eds.), *Spatial information theory: A theoretical basis for GIS (Proceedings of COSIT 1993)* (pp. 312–321). Berlin: Springer.

Montello, D. R. (1997). The perception and cognition of environmental distance: Direct sources of information. In S. C. Hirtle & A. U. Frank (Eds.), *Spatial information theory: A theoretical basis for GIS (Proceedings of COSIT 1997)* (pp. 297–311). Berlin: Springer.

Montello, D. R. (2005). Navigation. In P. Shah & A. Miyake (Eds.), *The Cambridge handbook of visuospatial thinking* (pp. 257–294). Cambridge, MA: Cambridge University Press.

Montello, D. R. (2009). A conceptual model of the cognitive processing of environmental distance information. In K. Stewart Hornsby, C. Claramunt, M. Denis, & G. Ligozat (Eds.), *Spatial information theory (Proceedings of COSIT 2009)* (pp. 1–17). Berlin: Springer.

Montello, D. R. (2014). Spatial cognition and architectural space: Research perspectives. In C. Derix & Å. Izaki (Eds.), Empathic space: The computation of human-centric architecture. *Architectural Design, 84*(5), 74–79.

Montello, D. R., Grossner, K., & Janelle, D. G. (Eds.) (2014). *Space in mind: Concepts for spatial learning and education*. Cambridge, MA: The MIT Press.

Montello, D. R., & Moyes, H. (2012). Why dark zones are sacred: Turning to behavioral and cognitive science for answers. In H. Moyes (Ed.), *Sacred darkness: A global perspective on the ritual use of caves* (pp. 385–396). Boulder, CO: University Press of Colorado.

Montello, D. R., & Raubal, M. (2013). Functions and applications of spatial cognition. In D. Waller & L. Nadel (Eds.), *Handbook of spatial cognition* (pp. 249–264). Washington, DC: American Psychological Association.

Morgan, L. K., MacEvoy, S. P., Aguirre, G. K., & Epstein, R. A. (2011). Distances between real-word locations are represented in the human hippocampus. *Journal of Neuroscience, 31*, 1238–1245.

Morganti, F., Stefanini, S., & Riva, G. (2013). From allo- to egocentric spatial ability in early Alzheimer's disease: A study with virtual reality spatial tasks. *Cognitive Neuroscience, 4*, 171–180.

Morrow, D. G., Bower, G. H., & Greenspan, S. L. (1989). Updating situation models during narrative comprehension. *Journal of Memory and Language, 28*, 292–312.

Morrow, D. G., & Clark, H. H. (1988). Interpreting words in spatial descriptions. *Language and Cognitive Processes, 3*, 275–291.

Morrow, D. G., Greenspan, S. L., & Bower, G. H. (1987). Accessibility and situation models in narrative comprehension. *Journal of Memory and Language, 26*, 165–187.

Moser, E. I., Kropff, E., & Moser, M.-B. (2008). Place cells, grid cells, and the brain's spatial representation system. *Annual Review of Neuroscience, 31*, 69–89.

Münzer, S., & Hölscher, C. (2011). Entwicklung und Validierung eines Fragebogens zu räumlichen Strategien [Development and validation of a questionnaire of spatial strategies]. *Diagnostica, 57*, 111–125.

Münzer, S., Zimmer, H. D., Schwalm, M., Baus, J., & Aslan, I. (2006). Computer-assisted navigation and the acquisition of route and survey knowledge. *Journal of Environmental Psychology, 26*, 300–308.

Nadel, L. (2013). Cognitive maps. In D. Waller & L. Nadel (Eds.), *Handbook of spatial cognition* (pp. 155–171). Washington, DC: American Psychological Association.

National Research Council (2006). *Learning to think spatially: GIS as a support system in the K-12 curriculum*. Washington, DC: National Academies Press.

Noordzij, M. L., & Postma, A. (2005). Categorical and metric distance information in mental representations derived from route and survey descriptions. *Psychological Research, 69*, 221–232.

Noordzij, M. L., Zuidhoek, S., & Postma, A. (2006). The influence of visual experience on the ability to form spatial mental models based on route and survey descriptions. *Cognition, 100*, 321–342.

Nothegger, C., Winter, S., & Raubal, M. (2004). Selection of salient features for route directions. *Spatial Cognition and Computation, 4*, 113–136.

Nys, M., Gyselinck, V., Orriols, E., & Hickmann, M. (2015). Landmark and route knowledge in children's spatial representation of a virtual environment. *Frontiers in Psychology, 5*: 1522.

O'Keefe, J., & Dostrovsky, J. (1971). The hippocampus as a spatial map: Preliminary evidence from unit activity in the freely-moving rat. *Brain Research, 34*, 171–175.

O'Keefe, J., & Nadel, L. (1978). *The hippocampus as a cognitive map*. Oxford: Oxford University Press.

Opheim, A., Jahnsen, R., Olsson, E., & Stanghelle, J. K. (2009). Walking function, pain, and fatigue in adults with cerebral palsy: A 7-year follow-up study. *Developmental Medicine and Child Neurology, 51*, 381–388.

Padgitt, A. J., & Hund, A. M. (2012). How good are these directions? Determining direction quality and wayfinding efficiency. *Journal of Environmental Psychology, 32*, 164–172.

Pailhous, J. (1970). *La représentation de l'espace urbain: L'exemple du chauffeur de taxi* [The representation of urban space: The example of taxi drivers]. Paris: Presses Universitaires de France.

Paillard, J. (Ed.) (1991). *Brain and space*. Oxford: Oxford University Press.

Palsky, G. (1999). Borges, Carroll et la carte au 1/1 [Borges, Carroll, and the 1:1 map]. *Cybergeo: European Journal of Geography*, Document 106.

Panofsky, E. (1991). *Perspective as symbolic form*. New York: Urzone.

Pantalides, S. N., Kelly, J. W., & Avraamides, M. N. (2016). Integration of spatial information across vision and language. *Journal of Cognitive Psychology, 28*, 171–185.

Parush, A., Ahuvia-Pick, S., & Erev, I. (2007). Degradation in spatial knowledge acquisition when using automatic navigation systems. In S. Winter, M. Duckham, L. Kulik, & B. Kuipers (Eds.), *Spatial information theory (Proceedings of COSIT 2007)* (pp. 238–254). Berlin: Springer.

Pasqualotto, A., & Proulx, M. J. (2012). The role of visual experience for the neural basis of spatial cognition. *Neuroscience and Biobehavioral Reviews, 36*, 1179–1187.

Pasqualotto, A., Spiller, M. J., Jansari, A. S., & Proulx, M. J. (2013). Visual experience facilitates allocentric representation. *Behavioural Brain Research, 236*, 175–179.

Passini, R., Rainville, C., Marchand, N., & Joanette, Y. (1995). Wayfinding in dementia of the Alzheimer type: Planning abilities. *Journal of Clinical and Experimental Neuropsychology, 17*, 820–832.

Pazzaglia, F., & Cornoldi, C. (1999). The role of distinct components of visuo-spatial working memory in the processing of texts. *Memory, 7*, 19–41.

Pazzaglia, F., De Beni, R., & Meneghetti, C. (2007). The effects of verbal and spatial interference in the encoding and retrieval of spatial and nonspatial texts. *Psychological Research, 71*, 484–494.

Pazzaglia, F., & Meneghetti, C. (2012). Spatial text processing in relation to spatial abilities and spatial styles. *Journal of Cognitive Psychology, 24*, 972–980.

Pazzaglia, F., Meneghetti, C., De Beni, R., & Gyselinck, V. (2010). Working memory components in survey and route spatial text processing. *Cognitive Processing, 11*, 359–369.

Pazzaglia, F., & Taylor, H. A. (2007). Perspective, instruction, and cognitive style in spatial representation of a virtual environment. *Spatial Cognition and Computation, 7*, 349–364.

Pederson, E. (1995). Language as context, language as means: Spatial cognition and habitual language use. *Cognitive Linguistics, 6*, 33–62.

Perec, G. (1997). *Species of spaces and other pieces*. London: Penguin.

Perec, G. (2010). *An attempt at exhausting a place in Paris*. Cambridge, MA: Wakefield Press.

Perrig, W., & Kintsch, W. (1985). Propositional and situational representations of text. *Journal of Memory and Language, 24*, 503–518.

Péruch, P., Chabanne, V., Nesa, M.-P., Thinus-Blanc, C., & Denis, M. (2006). Comparing distances in mental images constructed from visual experience or verbal descriptions: The impact of survey versus route perspective. *Quarterly Journal of Experimental Psychology, 59*, 1950–1967.

Péruch, P., Giraudo, M.-D., & Gärling, T. (1989). Distance cognition by taxi drivers and the general public. *Journal of Environmental Psychology, 9*, 233–239.

Péruch, P., & Latini Corazzini, L. (2006). Se déplacer et naviguer dans l'espace [Moving and navigating in space]. In A. Berthoz & J.-L. Vercher (Eds.), *Le traité de la réalité virtuelle*. Vol. 1: *L'homme et l'environnement virtuel* (pp. 189–208). Paris: Presses de l'Ecole des Mines.

Pesenti, M., Zago, L., Crivello, F., Mellet, E., Samson, D., Duroux, B., Seron, X., Mazoyer, B., & Tzourio-Mazoyer, N. (2001). Mental calculation in a prodigy is sustained by right prefrontal and medial temporal areas. *Nature Neuroscience*, *4*, 103–107.

Philbeck, J. W., Behrmann, M., Levy, L., Potolicchio, S. J., & Caputy, A. J. (2004). Path integration deficits during linear locomotion after human medial temporal lobectomy. *Journal of Cognitive Neuroscience*, *16*, 510–520.

Philbeck, J. W., & Sargent, J. (2013). Perception of spatial relations during self-motion. In D. Waller & L. Nadel (Eds.), *Handbook of spatial cognition* (pp. 99–115). Washington, DC: American Psychological Association.

Piaget, J., & Inhelder, B. (1967). *The child's conception of space*. New York: W. W. Norton.

Piaget, J., Inhelder, B., & Szeminska, A. (1960). *The child's conception of geometry*. New York: Basic Books.

Picinali, L., Afonso, A., Denis, M., & Katz, B. F. G. (2014). Exploration of architectural spaces by blind people using auditory virtual reality for the construction of spatial knowledge. *International Journal of Human-Computer Studies*, *72*, 393–407.

Pielot, M., & Boll, S. (2010). Tactile Wayfinder: Comparison of tactile waypoint navigation with commercial pedestrian navigation systems. In P. Floréen, A. Krüger, & M. Spasojevic (Eds.), *Pervasive Computing (Proceedings of the Eighth International Conference, Pervasive 2010)* (pp. 76–93). Berlin: Springer.

Pielot, M., Poppinga, B., Heuten, W., & Boll, S. (2012). PocketNavigator: Studying tactile navigation systems in-situ. In J. A. Konstan, E. H. Chi, & K. Höök (Eds.), *Proceedings of the SIGCHI Conference on Human Factors in Computing Systems* (pp. 3131–3140). New York: ACM.

Plancher, G., Gyselinck, V., Nicolas, S., & Piolino, P. (2010). Age effects on components of episodic memory and feature binding: A virtual reality study. *Neuropsychology*, *24*, 379–390.

Poincaré, H. (1905). *Science and hypothesis*. London: Walter Scott Publishing Co.

Poincaré, H. (1914). *Science and method*. London: Thomas Nelson and Sons.

Portugali, J. (1993). *Implicate relations: Society and space in the Israeli-Palestinian conflict*. Dordrecht, The Netherlands: Kluwer.

Postma, A., & van der Ham, I. J. M. (Eds.) (2017). *Neuropsychology of space: Spatial functions of the human brain*. London: Academic Press.

Poucet, B., Lenck-Santini, P. P., Hok, V., Save, E., Banquet, J.-P., Gaussier, P., & Muller, R. U. (2004). Spatial navigation and hippocampal place cell firing: The problem of goal encoding. *Reviews in the Neurosciences*, *15*, 89–107.

Pruden, S. M., Levine, S. C., & Huttenlocher, J. (2011). Children's spatial thinking: Does talk about the spatial world matter? *Developmental Science*, *14*, 1417–1430.

Pumain, D. (2016). Space and territory: Toward integrated concepts. In P. Beckouche, C. Grasland, F. Guérin-Pace, & J.-Y. Moisseron (Eds.), *Founding territorial sciences* (pp. 39–53). Paris: Collège International des Sciences du Territoire.

Raffestin, C. (1980). *Pour une géographie du pouvoir* [For a geography of power]. Paris: Litec.

Raffestin, C. (1986). Ecogenèse territoriale et territorialité [Territorial ecopoiesis and territoriality]. In F. Auriac & R. Brunet (Eds.), *Espace, jeux et enjeux* (pp. 175–185). Paris: Fayard/Fondation Diderot.

Raubal, M., & Egenhofer, M. J. (1998). Comparing the complexity of wayfinding tasks in built environments. *Environment and Planning B*, *25*, 895–913.

Raubal, M., & Winter, S. (2002). Enriching wayfinding instructions with local landmarks. In M. J. Egenhofer & D. M. Mark (Eds.), *Geographic information science* (pp. 243–259). Berlin: Springer.

Redish, A. D., & Ekstrom, A. (2013). Hippocampus and related areas: What the place cell literature tells us about cognitive maps in rats and humans. In D. Waller & L. Nadel (Eds.), *Handbook of spatial cognition* (pp. 15–34). Washington, DC: American Psychological Association.

Regian, J. W., & Yadrick, R. M. (1994). Assessment of configurational knowledge of naturally- and artificially-acquired large-scale space. *Journal of Environmental Psychology, 14*, 211–223.

Ricciardi, E., Bonino, D., Pellegrini, S., & Pietrini, P. (2014). Mind the blind brain to understand the sighted one! Is there a supramodal cortical functional architecture? *Neuroscience and Biobehavioral Reviews, 41*, 64–77.

Ricciardi, E., & Pietrini, P. (2011). New light from the dark: What blindness can teach us about brain function. *Current Opinion in Neurology, 24*, 357–363.

Richter, K.-F., Dara-Abrams, D., & Raubal, M. (2010). Navigating and learning with location based services: A user-centric design. In G. Gartner & Y. Li (Eds.), *Proceedings of the 7th International Symposium on LBS and TeleCartography* (pp. 261–276). Guangzhou, China: Spatial Information Research Center.

Richter, K.-F., & Duckham, M. (2008). Simplest instructions: Finding easy-to-describe routes for navigation. In T. J. Cova, H. J. Miller, K. Beard, A. U. Frank, & M. F. Goodchild (Eds.), *Geographic information science* (pp. 274–289). Berlin: Springer.

Richter, K.-F., Tomko, M., & Winter, S. (2008). A dialog-driven process of generating route directions. *Computers, Environment and Urban Systems, 32*, 233–245.

Richter, K.-F., & Winter, S. (2014). *Landmarks: GIScience for intelligent services*. Berlin: Springer.

Riesbeck, C. K. (1980). "You can't miss it!": Judging the clarity of directions. *Cognitive Science, 4*, 285–303.

Rieser, J. J. (1989). Access to knowledge of spatial structure at novel points of observation. *Journal of Experimental Psychology: Learning, Memory, and Cognition, 15*, 1157–1165.

Rieser, J. J., Guth, D. A., & Hill, E. W. (1986). Sensitivity to perspective structure while walking without vision. *Perception, 15*, 173–188.

Rinck, M., & Denis, M. (2004). The metrics of spatial distance traversed during mental imagery. *Journal of Experimental Psychology: Learning, Memory, and Cognition, 30*, 1211–1218.

Rinck, M., Hähnel, A., Bower, G. H., & Glowalla, U. (1997). The metrics of spatial situation models. *Journal of Experimental Psychology: Learning, Memory, and Cognition, 23*, 622–637.

Ritchie, K., Norton, J., Mann, A., Carrière, I., & Ancelin, M.-L. (2013). Late-onset agoraphobia: General population incidence and evidence for a clinical subtype. *American Journal of Psychiatry, 170*, 790–798.

Rivière, J., & Lécuyer, R. (2002). Spatial cognition in young children with spinal muscular atrophy. *Developmental Neuropsychology, 21*, 273–283.

Rivière, J., Lécuyer, R., & Hickmann, M. (2009). Early locomotion and the development of spatial language: Evidence from young children with motor impairments. *European Journal of Developmental Psychology, 6*, 548–566.

Robert, S. (2006). Deictic space in Wolof: Discourse, syntax and the importance of absence. In M. Hickmann & S. Robert (Eds.), *Space in languages: Linguistic systems and cognitive categories* (pp. 155–174). Amsterdam: John Benjamins.

Robin, F., & Denis, M. (1991). Description of perceived or imagined spatial networks. In R. H. Logie & M. Denis (Eds.), *Mental images in human cognition* (pp. 141–152). Amsterdam: North-Holland.

Rodrigo, M. J., de Vega, M., & Padrón, I. (2012). Common retrieval processes in producers and comprehenders of spatial information in a non-interactive task. *Journal of Cognitive Psychology, 24*, 342–358.

Rohmer, E. (2000). *L'organisation de l'espace dans le* Faust *de Murnau* [The organization of space in Murnau's *Faust*]. Paris: Ramsay.

Ruddle, R. A., & Lessels, S. (2009). The benefits of using a walking interface to navigate virtual environments. *ACM Transactions on Computer-Human Interaction, 16*, Article 5.

Ruddle, R. A., Payne, S. J., & Jones, D. M. (1997). Navigating buildings in "desk-top" virtual environments: Experimental investigations using extended navigational experience. *Journal of Experimental Psychology: Applied, 3*, 143–159.

Ruddle, R. A., Payne, S. J., & Jones, D. M. (1998). Navigating large-scale "desk-top" virtual buildings: Effects of orientation aids and familiarity. *Presence: Teleoperators and Virtual Environments*, 7, 179–192.

Ruddle, R. A., Payne, S. J., & Jones, D. M. (1999). Navigating large-scale virtual environments: What differences occur between helmet-mounted and desk-top displays? *Presence: Teleoperators and Virtual Environments, 8*, 157–168.

Ruddle, R. A., & Péruch, P. (2004). Effects of proprioceptive feedback and environmental characteristics on spatial learning in virtual environments. *International Journal of Human-Computer Studies, 60*, 299–326.

Ruddle, R. A., Volkova, E., & Bülthoff, H. H. (2011). Walking improves your cognitive map in environments that are large-scale and large in extent. *ACM Transactions on Computer-Human Interaction, 18*, Article 10.

Rusconi, E., Kwan, B., Giordano, B. L., Umiltà, C., & Butterworth, B. (2006). Spatial representation of pitch height: The SMARC effect. *Cognition, 99*, 113–129.

Sadalla, E. K., Burroughs, W. J., & Staplin, L. J. (1980). Reference points in spatial cognition. *Journal of Experimental Psychology: Human Learning and Memory, 6*, 516–528.

Sadalla, E. K., & Magel, S. G. (1980). The perception of traversed distance. *Environment and Behavior, 12*, 65–79.

Sadeghian, P., & Kantardzic, M. (2008). The new generation of automatic landmark detection systems: Challenges and guidelines. *Spatial Cognition and Computation, 8*, 252–287.

Saffiotti, A. (1997). The uses of fuzzy logic in autonomous robot navigation. *Soft Computing, 1*, 180–197.

Saint Aubert, E. de (2004). *Du lien des êtres aux éléments de l'être: Merleau-Ponty au tournant des années 1945–1951* [From the link of beings to the elements of beings: Merleau-Ponty at the turn of the years 1945–1951]. Paris: Vrin.

Schneider, L. F., & Taylor, H. A. (1999). How do you get there from here? Mental representations of route descriptions. *Applied Cognitive Psychology, 13*, 415–441.

Schober, M. F. (1993). Spatial perspective-taking in conversation. *Cognition, 47*, 1–24.

Schober, M. F. (2009). Spatial dialogue between partners with mismatched abilities. In K. R. Coventry, T. Tenbrink, & J. Bateman (Eds.), *Spatial language and dialogue* (pp. 23–29). Oxford: Oxford University Press.

Schuck, N. W., Doeller, C. F., Polk, T. A., Lindenberger, U., & Li, S.-C. (2015). Human aging alters the neural computation and representation of space. *NeuroImage, 117*, 141–150.

Seron, X., Pesenti, M., Noël, M.-P., Deloche, G., & Cornet, J.-A. (1992). Images of numbers, or "When 98 is upper left and 6 sky blue." *Cognition, 44*, 159–196.

Shah, P., & Miyake, A. (Eds.) (2005). *The Cambridge handbook of visuospatial thinking*. New York: Cambridge University Press.

Shanon, B. (1984). Room descriptions. *Discourse Processes, 7*, 225–255.

Shelton, A. L., & Gabrieli, J. D. E. (2002). Neural correlates of encoding space from route and survey perspectives. *Journal of Neuroscience, 22*, 2711–2717.

Shelton, A. L., & McNamara, T. P. (1997). Multiple views of spatial memory. *Psychonomic Bulletin and Review, 4*, 102–106.

Shelton, A. L., & McNamara, T. P. (2001). Systems of spatial reference in human memory. *Cognitive Psychology, 43*, 274–310.

Siegel, A. W., & White, S. H. (1975). The development of spatial representations of large-scale environments. In H. W. Reese (Ed.), *Advances in child development and behavior* (Vol. 10, pp. 9–55). New York: Academic Press.

Sorrows, M. E., & Hirtle, S. C. (1999). The nature of landmarks for real and electronic spaces. In C. Freksa & D. M. Mark (Eds.), *Spatial information theory: Cognitive and computational foundations of geographic information science (Proceedings of COSIT 1999)* (pp. 37–50). Berlin: Springer.

Spelke, E., Lee, S. A., & Izard, V. (2010). Beyond core knowledge: Natural geometry. *Cognitive Science, 34*, 863–884.

Spiers, H. J., & Maguire, E. A. (2006). Thoughts, behaviour, and brain dynamics during navigation in the real world. *NeuroImage, 31*, 1826–1840.

Spiers, H. J., Maguire, E. A., & Burgess, N. (2001). Hippocampal amnesia. *Neurocase, 7*, 357–382.

Staplin, L. J., & Sadalla, E. K. (1981). Distance cognition in urban environments. *The Professional Geographer, 33*, 302–310.

Stea, D., Blaut, J. M., & Stephens, J. (1996). Mapping as a cultural universal. In J. Portugali (Ed.), *The construction of cognitive maps* (pp. 345–360). Dordrecht, The Netherlands: Kluwer.

Steadman, P. (2001). *Vermeer's camera: Uncovering the truth behind the masterpieces*. Oxford: Oxford University Press.

Stevens, A., & Coupe, P. (1978). Distortions in judged spatial relations. *Cognitive Psychology, 10*, 422–437.

Streeter, L. A., Vitello, D., & Wonsiewicz, S. A. (1985). How to tell people where to go: Comparing navigational aids. *International Journal of Man-Machine Studies, 22*, 549–562.

Struiksma, M. E., Noordzij, M. L., Neggers, S. F. W., Bosker, W. M., & Postma, A. (2011). Spatial language processing in the blind: Evidence for a supramodal representation and cortical reorganization. *PLoS ONE, 6*, e24253.

Struiksma, M. E., Noordzij, M. L., & Postma, A. (2009). What is the link between language and spatial images? Behavioral and neural findings in blind and sighted individuals. *Acta Psychologica, 132*, 145–156.

Taillade, M., Sauzéon, H., Dejos, M., Arvind Pala, P., Larrue, F., Wallet, G., Gross, C., & N'Kaoua, B. (2013). Executive and memory correlates of age-related differences in wayfinding performances using a virtual reality application. *Aging, Neuropsychology, and Cognition, 20*, 298–319.

Tarry, G. (1886). Géométrie de situation: Nombre de manières distinctes de parcourir en une seule course toutes les allées d'un labyrinthe rentrant, en ne passant qu'une seule fois par chacune des allées [Geometry in situation: Number of distinct ways of traveling in a single course along all the paths of a maze by passing only once by each path]. In *Association Française pour l'Avancement des Sciences, Compte Rendu de la 15*ème *Session, Seconde Partie* (pp. 49–53). Paris: Association Française pour l'Avancement des Sciences.

Tarry, G. (1895). Le problème des labyrinthes [The maze problem]. *Nouvelles Annales de Mathématiques, 14*, 187–190.

Taylor, H. A., & Tversky, B. (1992). Spatial mental models derived from survey and route descriptions. *Journal of Memory and Language, 31*, 261–292.

Taylor, H. A., & Tversky, B. (1996). Perspective in spatial descriptions. *Journal of Memory and Language, 35*, 371–391.

Techentin, C., Voyer, D., & Voyer, S. D. (2014). Spatial abilities and aging: A meta-analysis. *Experimental Aging Research, 40*, 395–425.

Tenbrink, T., Wiener, J., & Claramunt, C. (2013). Representing space in cognitive science: From empirical insights via computational models to human-centred assistance. In T. Tenbrink, J. Wiener, & C. Claramunt (Eds.), *Representing space in cognition: Interrelations of behaviour, language, and formal models* (pp. 1–7). Oxford: Oxford University Press.

Tenbrink, T., & Winter, S. (2009). Variable granularity in route directions. *Spatial Cognition and Computation, 9*, 64–93.

Tezuka, T., & Tanaka, K. (2005). Landmark extraction: A web mining approach. In A. G. Cohn & D. M. Mark (Eds.), *Spatial information theory (Proceedings of COSIT 2005)* (pp. 379–396). Berlin: Springer.

Thinus-Blanc, C., & Gaunet, F. (1997). Representation of space in blind persons: Vision as a spatial sense? *Psychological Bulletin, 121*, 20–42.

Thomson, J. A. (1983). Is continuous visual monitoring necessary in visually guided locomotion? *Journal of Experimental Psychology: Human Perception and Performance, 9*, 427–443.

Thorndyke, P. W. (1981). Distance estimation from cognitive maps. *Cognitive Psychology, 13*, 526–550.

Thorndyke, P. W., & Hayes-Roth, B. (1982). Differences in spatial knowledge acquired from maps and navigation. *Cognitive Psychology, 14*, 560–589.

Thrift, N. (2003). Space: The fundamental stuff of human geography. In S. Holloway, S. Rice, & G. Valentine (Eds.), *Key concepts in geography* (pp. 95–107). London: SAGE.

Thrun, S. (2003). Robotic mapping: A survey. In G. Lakemeyer & B. Nebel (Eds.), *Exploring artificial intelligence in the new millennium* (pp. 1–35). San Francisco, CA: Morgan Kaufmann.

Tinti, C., Adenzato, M., Tamietto, M., & Cornoldi, C. (2006). Visual experience is not necessary for efficient survey spatial cognition: Evidence from blindness. *Quarterly Journal of Experimental Psychology, 59*, 1306–1328.

Tolman, E. C. (1948). Cognitive maps in rats and men. *Psychological Review, 55*, 189–208.

Tom, A., & Denis, M. (2003). Referring to landmark or street information in route directions: What difference does it make? In W. Kuhn, M. F. Worboys, & S. Timpf (Eds.), *Spatial information theory: Foundations of geographic information science (Proceedings of COSIT 2003)* (pp. 384–397). Berlin: Springer.

Tom, A., & Denis, M. (2004). Language and spatial cognition: Comparing the roles of landmarks and street names in route instructions. *Applied Cognitive Psychology, 18*, 1213–1230.

Tom, A. C., & Tversky, B. (2012). Remembering routes: Streets and landmarks. *Applied Cognitive Psychology, 26*, 182–193.

Tomko, M., & Richter, K.-F. (2015). Defensive wayfinding: Incongruent information in route following. In S. I. Fabrikant, M. Raubal, M. Bertolotto, C. Davies, S. Freundschuh, & S. Bell (Eds.), *Spatial information theory (Proceedings of COSIT 2015)* (pp. 426–446). Cham, Switzerland: Springer.

Tversky, B. (1981). Distortions in memory for maps. *Cognitive Psychology, 13*, 407–433.

Tversky, B. (1993). Cognitive maps, cognitive collages, and spatial mental models. In A. U. Frank & I. Campari (Eds.), *Spatial information theory: A theoretical basis for GIS* (pp. 14–24). Berlin: Springer.

Tversky, B. (2003). Structures on mental spaces: How people think about space. *Environment and Behavior, 35*, 66–80.

Tversky, B. (2005). Functional significance of visuospatial representations. In P. Shah & A. Miyake (Eds.), *The Cambridge handbook of visuospatial thinking* (pp. 1–34). Cambridge, MA: Cambridge University Press.

Tversky, B., & Lee, P. U. (1998). How space structures language. In C. Freksa, C. Habel, & K. F. Wender (Eds.), *Spatial cognition: An interdisciplinary approach to representing and processing spatial knowledge* (pp. 157–175). Berlin: Springer.

Tversky, B., & Lee, P. U. (1999). Pictorial and verbal tools for conveying routes. In C. Freksa & D. M. Mark (Eds.), *Spatial information theory: Cognitive and computational foundations of geographic information science (Proceedings of COSIT 1999)* (pp. 51–64). Berlin: Springer.

Tversky, B., Taylor, H. A., & Mainwaring, S. (1997). Langage et perspective spatiale [Language and spatial perspective]. In M. Denis (Ed.), *Langage et cognition spatiale* (pp. 25–49). Paris: Masson.

Uc, E. Y., Rizzo, M., Anderson, S. W., Sparks, J. D., Rodnitzky, R. L., & Dawson, J. D. (2007). Impaired navigation in drivers with Parkinson's disease. *Brain, 130*, 2433–2440.

Ungerleider, L. G., & Mishkin, M. (1982). Two cortical visual systems. In D. J. Ingle, M. A. Goodale, & R. J. W. Mansfield (Eds.), *Analysis of visual behavior* (pp. 549–586). Cambridge, MA: The MIT Press.

Vandeloise, C. (1986). *L'espace en français: Sémantique des prépositions spatiales* [Space in French: Semantics of spatial prepositions]. Paris: Editions du Seuil.

Vandeloise, C. (1991). *Spatial prepositions: A case study in French*. Chicago, IL: University of Chicago Press.

Vandeloise, C. (2008). Three basic prepositions in French and in English: A comparison. *Carnets de Grammaire*, No. 19. Toulouse, France: CLLE-ERSS.

van der Ham, I. J. M., Kant, N., Postma, A., & Visser-Meily, J. M. A. (2013). Is navigation ability a problem in mild stroke patients? Insights from self-reported navigation measures. *Journal of Rehabilitation Medicine, 45*, 429–433.

van Dijk, T. A., & Kintsch, W. (1983). *Strategies of discourse comprehension*. New York: Academic Press.

Vanetti, E. J., & Allen, G. L. (1988). Communicating environmental knowledge: The impact of verbal and spatial abilities on the production and comprehension of route directions. *Environment and Behavior, 20*, 667–682.

Veraart, C., & Wanet-Defalque, M.-C. (1987). Representation of locomotor space by the blind. *Perception and Psychophysics, 42*, 132–139.

Vernant, J.-P. (1965). *Mythe et pensée chez les Grecs: Etudes de psychologie historique* [Myth and thought in Greece: Studies of historical psychology]. Paris: François Maspero.

Vetö, M. (2008). L'eidétique de l'espace chez Merleau-Ponty [Merleau-Ponty's eidetics of space]. *Archives de Philosophie, 71*, 407–438.

Viaud-Delmon, I., Berthoz, A., & Jouvent, R. (2002). Multisensory integration for spatial orientation in trait anxiety subjects: Absence of visual dependence. *European Psychiatry, 17*, 194–199.

Vidal, M., Amorim, M.-A., & Berthoz, A. (2004). Navigating in a virtual three-dimensional maze: How do egocentric and allocentric reference frames interact? *Cognitive Brain Research, 19*, 244–258.

Wagener, M., & Wender, K. F. (1985). Spatial representations and inference processes in memory for text. In G. Rickheit & H. Strohner (Eds.), *Inferences in text processing* (pp. 115–136). Amsterdam: North-Holland.

Waller, D. (1999). Factors affecting the perception of interobject distances in virtual environments. *Presence: Teleoperators and Virtual Environments, 8*, 657–670.

Waller, D., Beall, A. C., & Loomis, J. M. (2004). Using virtual environments to assess directional knowledge. *Journal of Environmental Psychology, 24*, 105–116.

Waller, D., Hunt, E., & Knapp, D. (1998). The transfer of spatial knowledge in virtual environment training. *Presence: Teleoperators and Virtual Environments*, 7, 129–143.

Waller, D., & Lippa, Y. (2007). Landmarks as beacons and associative cues: Their role in route learning. *Memory and Cognition, 35*, 910–924.

Waller, D., & Nadel, L. (Eds.) (2013). *Handbook of spatial cognition*. Washington, DC: American Psychological Association.

Waller, G. (1986). Linear organization of spatial instructions: Development of comprehension and production. *First Language, 6*, 53–67.

Wallgrün, J. O. (2010). *Hierarchical Voronoi graphs: Spatial representation and reasoning for mobile robots*. Berlin: Springer.

Walsh, V. (2003). A theory of magnitude: Common cortical metrics of time, space and quantity. *Trends in Cognitive Sciences*, 7, 483–488.

Wang, L., Mou, W., & Sun, X. (2014). Development of landmark knowledge at decision points. *Spatial Cognition and Computation, 14*, 1–17.

Wang, R. F. (2004). Between reality and imagination: When is spatial updating automatic? *Perception and Psychophysics, 66*, 68–76.

Wassmann, J., & Dasen, P. R. (1998). Balinese spatial orientation: Some empirical evidence of moderate linguistic relativity. *Journal of the Royal Anthropological Institute, 4*, 689–711.

Wegman, J., & Janzen, G. (2011). Neural encoding of objects relevant for navigation and resting state correlations with navigational ability. *Journal of Cognitive Neuroscience, 23*, 3841–3854.

Wehner, R. (2003). Desert ant navigation: How miniature brains solve complex tasks. *Journal of Comparative Physiology A, 189*, 579–588.

Wehner, R., & Wehner, S. (1990). Insect navigation: Use of maps or Ariadne's thread? *Ethology Ecology and Evolution, 2*, 27–48.

Whorf, B. L. (1956). *Language, thought, and reality: Selected writings of Benjamin Lee Whorf* (J. B. Carroll, Ed.). Cambridge, MA: The MIT Press.

Wiener, J. M., Büchner, S. J., & Hölscher, C. (2009). Taxonomy of human wayfinding tasks: A knowledge-based approach. *Spatial Cognition and Computation, 9*, 152–165.

Wiener, J. M., & Mallot, H. A. (2003). "Fine-to-coarse" route planning and navigation in regionalized environments. *Spatial Cognition and Computation, 3*, 331–358.

Wiener, J. M., & Mallot, H. A. (2006). Path complexity does not impair visual path integration. *Spatial Cognition and Computation, 6*, 333–346.

Wilkniss, S. M., Jones, M. G., Korol, D. L., Gold, P. E., & Manning, C. A. (1997). Age-related differences in an ecologically based study of route learning. *Psychology and Aging, 12*, 372–375.

Wilton, R. N. (1979). Knowledge of spatial relations: The specification of the information used in making inferences. *Quarterly Journal of Experimental Psychology, 31*, 133–146.

Wineman, J. D., & Peponis, J. (2010). Constructing spatial meaning: Spatial affordances in museum design. *Environment and Behavior, 42*, 86–109.

Winock, M. (2013). *Flaubert*. Paris: Gallimard.

Winter, S. (2009). Spatial intelligence: Ready for a challenge? *Spatial Cognition and Computation, 9*, 138–151.

Wolbers, T., & Büchel, C. (2005). Dissociable retrosplenial and hippocampal contributions to successful formation of survey representations. *Journal of Neuroscience, 25*, 3333–3340.

Wolbers, T., & Hegarty, M. (2010). What determines our navigational abilities? *Trends in Cognitive Sciences, 14*, 138–146.

Wolbers, T., & Wiener, J. M. (2014). Challenges for identifying the neural mechanisms that support spatial navigation: The impact of spatial scale. *Frontiers in Human Neuroscience, 8*: 571.

Wolter, D. (2008). *Spatial representation and reasoning for robot mapping: A shape-based approach*, Berlin: Springer.

Wood, G., Willmes, K., Nuerk, H.-C., & Fischer, M. H. (2008). On the cognitive link between space and number: A meta-analysis of the SNARC effect. *Psychology Science Quarterly, 50*, 489–525.

Woollett, K., Spiers, H. J., & Maguire, E. A. (2009). Talent in the taxi: A model system for exploring expertise. *Philosophical Transactions of the Royal Society of London B, 364*, 1407–1416.

Worsley, C. L., Recce, M., Spiers, H. J., Marley, J., Polkey, C. E., & Morris, R. G. (2001). Path integration following temporal lobectomy in humans. *Neuropsychologia, 39*, 452–464.

Wunderlich, D., & Reinelt, R. (1982). How to get there from here. In R. J. Jarvella & W. Klein (Eds.), *Speech, place, and action: Studies in deixis and related topics* (pp. 183–201). Chichester, UK: Wiley.

Xenakis, I. (1971). *Formalized music: Thought and mathematics in composition.* Bloomington, IN: Indiana University Press.

Zacks, J., Rypma, B., Gabrieli, J. D. E., Tversky, B., & Glover, G. H. (1999). Imagined transformations of bodies: An fMRI investigation. *Neuropsychologia, 37*, 1029–1040.

Zanbaka, C., Lok, B., Babu, S., Ulinski, A., & Hodges, L. F. (2005). Comparison of path visualizations and cognitive measures relative to travel technique in a virtual environment. *IEEE Transactions on Visualization and Computer Graphics, 11*, 694–705.

Zancada-Menéndez, C., He, Q., Sampedro-Piquero, P., Lopez, L., & McNamara, T. P. (2016). Influence of bidirectional perspective on learning routes and spatial layout. *Journal of Cognitive Psychology, 28*, 474–485.

Zancada-Menéndez, C., Sampedro-Piquero, P., Lopez, L., & McNamara, T. P. (2016). Age and gender differences in spatial perspective taking. *Aging Clinical and Experimental Research, 28*, 289–296.

Zevi, B. (1957). *Architecture as space: How to look at architecture.* New York: Horizon Press.

INDEX OF NAMES

INDEX OF TERMS